Master Math: Basic Math and Pre-Algebra

By

Debra Anne Ross

Course Technology PTR
A part of Cengage Learning

Australia • Brazil • Japan • Korea • Mexico • Singapore • Spain • United Kingdom • United States

Publisher and General Manager, Course Technology PTR:
Stacy L. Hiquet

Associate Director of Marketing:
Sarah Panella

Manager of Editorial Services:
Heather Talbot

Marketing Manager: Jordan Casey

Senior Acquisitions Editor:
Emi Smith

Interior Layout Tech:
Judith Littlefield

Illustrations and Equations:
Judith Littlefield

Cover Designer: Jeff Cooper

Indexer: Larry Sweazy

Proofreader: Jenny Davidson

Library of Congress Control Number: 2009924539

ISBN-10: 1-59863-982-X
ISBN-13: 978-1-59863-982-7

Course Technology, a part of Cengage Learning
20 Channel Center Street
Boston, MA 02210
USA

Cengage Learning is a leading provider of customized learning solutions with office locations around the globe, including Singapore, the United Kingdom, Australia, Mexico, Brazil, and Japan. Locate your local office at: **international.cengage.com/region**

Cengage Learning products are represented in Canada by Nelson Education, Ltd. For your lifelong learning solutions, visit **courseptr.com**
Visit our corporate website at **cengage.com**

Printed in Canada
1 2 3 4 5 6 7 12 11 10 09

Table of Contents

Acknowledgments

I sincerely thank Dr. Melanie McNeil, Professor of Chemical Engineering at San Jose State University, for reading this book for accuracy and for all her helpful comments. I am grateful to Dr. Channing Robertson, Professor of Chemical Engineering at Stanford University, for reviewing this book and, in general, for his sagacious guidance. I especially thank my mother, Maggie Ross, for reading this book and for her editorial help.

Without my wonderful agent, Sidney B. Kramer, and the staff of Mews Books, the *Master Math* series would not have been published. Thank you, Sidney! I am also thankful to Ron Fry and the staff of Career Press for their work in publishing and launching the original *Master Math* books as a successful series.

I am grateful to Emi Smith, Acquisitions Editor, and Course Technology, a part of Cengage Learning, for invigorating the *Master Math* series and improving the presentation. I particularly appreciate Judith Littlefield for her tireless and expert work on the illustrations, equations, and layout. Much thanks to Jenny Davidson for proofreading, Jeff Cooper for cover design, Larry Sweazy for indexing, as well as Stacy L. Hiquet, Sarah Panella, Heather Talbot, and Jordan Casey.

Finally, I deeply appreciate my beautiful and brilliant husband, David A. Lawrence, who worked side-by-side with me as we meticulously edited text and figures.

About the Author

Debra Anne Ross Lawrence is the author of six books of the *Master Math* series: *Basic Math and Pre-Algebra*, *Algebra*, *Pre-Calculus*, *Calculus*, *Trigonometry*, and *Geometry*. She earned a double Bachelor of Arts degree in biology and chemistry with honors from the University of California at Santa Cruz and a Master of Science degree in chemical engineering from Stanford University.

Her research experience encompasses investigating the photosynthetic light reactions using a dye laser, studying the eye lens of diabetic patients, creating a computer simulation program of physiological responses to sensory and chemical disturbances, genetically engineering bacteria cells for over-expression of a protein, and designing and fabricating biological reactors for in-vivo study of microbial metabolism using nuclear magnetic resonance spectroscopy.

Debra was a member of a small team of scientists and engineers who developed and brought to market the first commercial biosensor system. She managed an engineering group responsible for scale-up of combinatorial synthesis for pharmaceutical development. She also managed intellectual property for a scientific research and development company. Debra's work has been published in scientific journals and/or patented.

Debra is also the author of *The 3:00 PM Secret: Live Slim and Strong Live Your Dreams* and *The 3:00 PM Secret 10-Day Dream Diet*. She is the coauthor with her husband, David A. Lawrence, of *Arrows Through Time: A Time Travel Tale of Adventure, Courage, and Faith*. Debra is President of GlacierDog Publishing and Founder of GlacierDog.com. When Debra is not engaged in all-season mountaineering near her Alaska home, she is endeavoring to understand the incomprehensible workings of the universe.

Introduction

Basic Math and Pre-Algebra is the first book in the *Master Math* series. The series also includes *Algebra, Pre-Calculus, Geometry, Trigonometry,* and *Calculus*. The *Master Math* series presents the general principles of mathematics from grade school through college including arithmetic, algebra, geometry, trigonometry, pre-calculus, and introductory calculus.

Basic Math and Pre-Algebra is a comprehensive arithmetic book that explains the subject matter in a way that makes sense to the reader. It begins with the most basic fundamental principles and progresses through more advanced topics to prepare a student for algebra. *Basic Math and Pre-Algebra* explains the principles and operations of arithmetic, provides step-by-step procedures and solutions and presents examples and applications.

Basic Math and Pre-Algebra is a reference book for grade school and middle school students that explains and clarifies the arithmetic principles they are learning in school. It is also a comprehensive reference source for more advanced students learning algebra and pre-calculus. *Basic Math and Pre-Algebra* is invaluable for students of all ages, parents, tutors, and anyone needing a basic arithmetic reference source.

The information provided in each book and in the series as a whole is progressive in difficulty and builds on itself, which allows the reader to gain perspective on the connected nature of mathematics. The skills required to understand every topic presented are explained in an earlier chapter or book within the series. Each book contains a complete table of contents and a comprehensive index, so that specific subjects, principles and formulas can be easily found. The books are written in a simple style that facilitates understanding and easy referencing of sought-after principles, definitions, and explanations.

Basic Math and Pre-Algebra and the *Master Math* series are not replacements for textbooks but rather reference books providing explanations and perspective. The *Master Math* series would have been invaluable to me during my entire education from grade school through graduate school. There is no other source that provides the breadth and depth of the *Master Math* series in a single book or series.

Finally, mathematics is a language—the universal language. A person struggling with mathematics should approach it in the same fashion or he or she would approach learning any other language. If someone moves to a foreign country, he or she does not expect to know the language automatically. It takes practice and contact with a language in order to master it. After a short time in the foreign country he or she would not say, "I do not know this language well yet. I must not have an aptitude for it." Yet many people have this attitude toward mathematics. If time is spent learning and practicing the principles, mathematics will become familiar and understandable. Don't give up.

Chapter 1

Numbers and Their Operations

1.1 Digits and the Base Ten System

• The system of numbers used for counting is based on groups of ten and is called the base ten system.

• Numbers are made up of digits that each correspond to a value. For example, the number 5,639,248 or five million, six hundred thirty-nine thousand, two hundred forty-eight represents:

> 5 millions, 6 hundred thousands, 3 ten thousands, 9 thousands, 2 hundreds, 4 tens, and 8 ones

This number can also be written:

$$5{,}000{,}000 + 600{,}000 + 30{,}000 + 9{,}000 + 200 + 40 + 8$$

- The ones digit indicates the number of ones, (1).
- The tens digit indicates the number of tens, (10).
- The hundreds digit indicates the number of hundreds, (100).
- The thousands digit indicates the number of thousands, (1,000).
- The ten thousands digit indicates the number of tens of thousands, (10,000).
- The hundred thousands digit indicates the number of hundreds of thousands, (100,000).
- The millions digit indicates the number of millions, (1,000,000).

1.2 Whole Numbers

• Whole numbers include zero and the counting numbers greater than zero.

• Negative numbers and numbers in the form of fractions, decimals, percents, or exponents are *not* whole numbers.

• Whole numbers are depicted on the number line and include zero and numbers to the right of zero.

Whole Numbers =

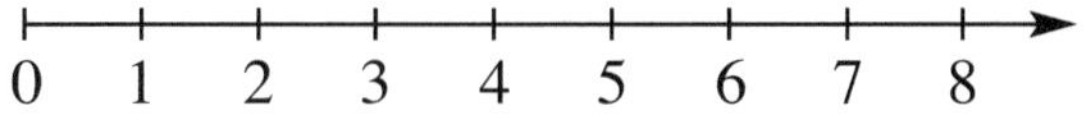

• Whole numbers can be written in the form of a set.

Whole Numbers = (0, 1, 2, 3, 4, 5, 6, 7, 8, 9, 10, 11, ...)

1.3 Addition of Whole Numbers

This section includes definitions, a detailed explanation of addition, and a detailed description of the standard addition technique.

• The symbol for addition is $+$.

• Addition is written $6 + 2$.

• The numbers to be added are called *addends*.

• The answer obtained in addition is called the *sum*.

• The symbol for what the sum is equal to is the equal sign $=$.

• Note: The symbol for not equal is $\neq$.

• For any number n, the following is true: $n + 0 = n$
(Note that letters are often used to represent numbers.)

A Detailed Explanation of Addition

- Consider the following examples:

$5 + 3 = 8$

$34 + 76 = 110$

$439 + 278 = 717$

$5 + 3 + 2 = 10$

$33 + 28 + 56 = 117$

$456 + 235 + 649 = 1{,}340$

- To add numbers that have two or more digits, it is easier to write the numbers in a column format with ones, tens, hundreds, etc., aligned, then add each column.

- **Example:** Add 5 + 3.

$$\begin{array}{r} 5 \text{ ones} \\ +\ 3 \text{ ones} \\ \hline 8 \text{ ones} \end{array}$$

- **Example:** Add 4 + 6.

$$\begin{array}{r} 4 \text{ ones} \\ +\ \ 6 \text{ ones} \\ \hline 10 \text{ ones} \end{array}$$

- **Example:** Add 7 + 0.

$$\begin{array}{r} 7 \text{ ones} \\ +\ 0 \text{ ones} \\ \hline 7 \text{ ones} \end{array}$$

- **Example:** Add $7 + 8$.

$$\begin{array}{r} 7 \text{ ones} \\ + \quad 8 \text{ ones} \\ \hline 15 \text{ ones} \end{array}$$

Note that 15 ones is equivalent to 1 ten and 5 ones.

- **Example:** Add $21 + 32$.

$$\begin{array}{r} 21 \\ + \ 32 \\ \hline ? \end{array}$$

This equation can be rewritten as:

$$\begin{array}{rrcr} & 2 \text{ tens} & + & 1 \text{ ones} \\ + & 3 \text{ tens} & + & 2 \text{ ones} \\ \hline & 5 \text{ tens} & + & 3 \text{ ones} \end{array}$$

$= 50 + 3 = 53$

- **Example:** Add $34 + 76$.

$$\begin{array}{r} 34 \\ + \ 76 \\ \hline ? \end{array}$$

This equation can be rewritten as:

$$\begin{array}{rrcr} & 3 \text{ tens} & + & 4 \text{ ones} \\ + & 7 \text{ tens} & + & 6 \text{ ones} \\ \hline & 10 \text{ tens} & + & 10 \text{ ones} \end{array}$$

$= 100 + 10 = 110$

This equation can also be rewritten as:

$$\begin{array}{r} 30 + \ \ 4 \\ + \ 70 + \ \ 6 \\ \hline 10 + 10 \end{array}$$

$= 110$

- **Example:** Add 7 + 28.

$$\begin{array}{r} 7 \\ + \ 28 \\ \hline ? \end{array}$$

This equation can be rewritten as:

$$\begin{array}{rrr} & 0 \text{ tens} + & 7 \text{ ones} \\ + & 2 \text{ tens} + & 8 \text{ ones} \\ \hline & 2 \text{ tens} + & 15 \text{ ones} \end{array}$$

$= 20 + 15 = 35$

- **Example:** Add 439 + 278.

$$\begin{array}{r} 439 \\ + \ 278 \\ \hline ? \end{array}$$

This equation can be rewritten as:

$$\begin{array}{rrrr} & 4 \text{ hundreds} + & 3 \text{ tens} + & 9 \text{ ones} \\ + & 2 \text{ hundreds} + & 7 \text{ tens} + & 8 \text{ ones} \\ \hline & 6 \text{ hundreds} + & 10 \text{ tens} + & 17 \text{ ones} \end{array}$$

$= 600 + 100 + 17 = 717$

This equation can also be rewritten as:

$$\begin{array}{rcrcr} 400 & + & 30 & + & 9 \\ +\ 200 & + & 70 & + & 8 \\ \hline 600 & + & 100 & + & 17 \end{array}$$

$= 717$

A Detailed Description of the Standard Addition Technique

- An alternative to separating the ones, tens, hundreds, etc., is to align the proper digits in columns (ones over ones, tens over tens, etc.). Then add each column, beginning with the ones, and carry-over digits to the left. (Carry over the tens generated in the ones' column into the tens' column, carry over the hundreds generated in the tens' column into the hundreds' column, carry over the thousands generated in the hundreds' column into the thousands' columns, and so on.)

- **Example:** Add 35 + 46.

$$\begin{array}{r} 35 \\ +\ 46 \\ \hline ? \end{array}$$

First add $5 + 6 = 11$, and carry over the 10 into the tens' column.

$$\begin{array}{r} \mathit{1} \\ 35 \\ +\ 46 \\ \hline 1 \end{array}$$

Add the tens' column.

$$\begin{array}{r} \mathit{1} \\ 35 \\ +\ 46 \\ \hline 81 \end{array}$$

- **Example:** Add 56 + 789

$$\begin{array}{r} 56 \\ +\ 789 \\ \hline ? \end{array} \qquad \text{Switching order:} \qquad \begin{array}{r} 789 \\ +\ \ 56 \\ \hline ? \end{array}$$

Add 9 + 6 = 15, and carry over the 10 into the tens' column.

$$\begin{array}{r} \mathit{1} \\ 789 \\ +\ \ 56 \\ \hline 5 \end{array}$$

Add the tens' column, 1 + 8 + 5 = 14 tens, and carry over 100 to the hundreds' column.

$$\begin{array}{r} \mathit{11} \\ 789 \\ +\ \ 56 \\ \hline 45 \end{array}$$

Add the hundreds' column, 1 + 7 = 8 hundreds.

$$\begin{array}{r} \mathit{11} \\ 789 \\ +\ \ 56 \\ \hline 845 \end{array}$$

- **Example:** Add 5 + 49 + 387 + 928.

$$\begin{array}{r} 928 \\ 387 \\ 49 \\ +\quad 5 \\ \hline ? \end{array}$$

Add the ones, 8 + 7 + 9 + 5 = 29, and carry over the 20 into the tens' column.

$$\begin{array}{r} \mathit{2} \\ 928 \\ 387 \\ 49 \\ +\quad 5 \\ \hline 9 \end{array}$$

Add the tens, 2 + 2 + 8 + 4 = 16 tens, and carry over the 100 into the hundreds' column.

$$\begin{array}{r} \mathit{12} \\ 928 \\ 387 \\ 49 \\ +\quad 5 \\ \hline 69 \end{array}$$

Add the hundreds, $1 + 9 + 3 = 13$ hundreds.

$$\begin{array}{r} \textit{12} \\ 928 \\ 387 \\ 49 \\ +\quad 5 \\ \hline 1369 \end{array}$$

1.4 Subtraction of Whole Numbers

This section includes definitions, a detailed explanation of subtraction, and a detailed description of the standard subtraction technique.

- Subtraction is the process of finding the difference between two numbers.
- The symbol for subtraction is –.
- Subtraction is written $6 - 2$.
- The number to be subtracted from is called the *minuend*.
- The number to be subtracted is the *subtrahend*.
- The answer obtained in subtraction is called the *difference*.
- Subtraction is the reverse of addition.

  If $2 + 3 = 5$, then $5 - 2 = 3$ or $5 - 3 = 2$.

- For any number n, the following is true: $n - 0 = n$

(Note that letters are often used to represent numbers.)

- To subtract more than two numbers, subtract the first two, then subtract the third number from the difference of the first two, and so on. Subtraction must be performed in the order that the numbers are listed.

A Detailed Explanation of Subtraction

- Consider the examples:

$6 - 2 = 4$

$26 - 8 = 18$

$10{,}322 - 899 = 9{,}423$

- To subtract numbers with two or more digits, it is easier to write the numbers in a column format with ones, tens, hundreds, etc., aligned, then subtract each column beginning with the ones. To subtract a digit in a given column from a smaller digit in the same column, borrow from the next larger column.

- **Example:** Subtract 6 – 2.

$$\begin{array}{r} 6 \text{ ones} \\ -\ 2 \text{ ones} \\ \hline 4 \text{ ones} \end{array}$$

- **Example:** Subtract 26 – 8.

$$\begin{array}{r} 26 \\ -\ \ 8 \\ \hline ? \end{array}$$

This equation can be rewritten as:

$$\begin{array}{rrr} & 2 \text{ tens} + & 6 \text{ ones} \\ - & 0 \text{ tens} + & 8 \text{ ones} \\ \hline & ? & \end{array}$$

Borrow 10 from the tens' column.

$$\begin{array}{rrr} & 1 \text{ tens} + & 16 \text{ ones} \\ - & 0 \text{ tens} + & 8 \text{ ones} \\ \hline & ? & \end{array}$$

Subtract the ones, $16 - 8 = 8$.

$$\begin{array}{rrr} & 1 \text{ tens} + & 16 \text{ ones} \\ - & 0 \text{ tens} + & 8 \text{ ones} \\ \hline & & 8 \text{ ones} \end{array}$$

Subtract the tens, $1 - 0 = 1$ ten.

$$\begin{array}{rrr} & 1 \text{ tens} + & 16 \text{ ones} \\ - & 0 \text{ tens} + & 8 \text{ ones} \\ \hline & 1 \text{ tens} + & 8 \text{ ones} \end{array}$$

$= 10 + 8 = 18$

- **Example:** Subtract $10{,}322 - 899$.

$$\begin{array}{rr} & 10{,}322 \\ - & 899 \\ \hline & ? \end{array}$$

$$\begin{array}{rrrrrr} & 1 \text{ ten thousands} + & 0 \text{ thousands} + & 3 \text{ hundreds} + & 2 \text{ tens} + & 2 \text{ ones} \\ - & 0 \text{ ten thousands} + & 0 \text{ thousands} + & 8 \text{ hundreds} + & 9 \text{ tens} + & 9 \text{ ones} \\ \hline & & & ? & & \end{array}$$

To subtract the ones, borrow 10 from the tens' column, then subtract 12 – 9 = 3.

	1 ten thousands	+	0 thousands	+	3 hundreds	+	1 tens	+	12 ones
–	0 ten thousands	+	0 thousands	+	8 hundreds	+	9 tens	+	9 ones
									3 ones

To subtract the tens, borrow 100 from the hundreds' column, then subtract 110 – 90 = 20.

	1 ten thousands	+	0 thousands	+	2 hundreds	+	11 tens	+	12 ones
–	0 ten thousands	+	0 thousands	+	8 hundreds	+	9 tens	+	9 ones
							2 tens	+	3 ones

To subtract the hundreds, first borrow 10,000 from the ten thousands' column into the thousands' column, then borrow 1,000 from the thousands' column.

	0 ten thousands	+	9 thousands	+	12 hundreds	+	11 tens	+	12 ones
–	0 ten thousands	+	0 thousands	+	8 hundreds	+	9 tens	+	9 ones
							2 tens	+	3 ones

Next, subtract the hundreds, 1,200 – 800 = 400.
Then, subtract the thousands, 9,000 – 0 = 9,000.
Then, subtract the ten thousands, 0 – 0 = 0.

	0 ten thousands	+	9 thousands	+	12 hundreds	+	11 tens	+	12 ones
–	0 ten thousands	+	0 thousands	+	8 hundreds	+	9 tens	+	9 ones
			9 thousands	+	4 hundreds	+	2 tens	+	3 ones

= 9,000 + 400 + 20 + 3 = 9,423

A Detailed Description of the Standard Subtraction Technique

• An alternative to separating the ones, tens, hundreds, etc., is to align the proper digits in columns and borrow (if needed) from the column to the left, which is 10 times larger.

• **Example:** Subtract 26 – 8.

$$\begin{array}{r} 26 \\ -\ \ 8 \\ \hline ? \end{array}$$

This equation can be rewritten as:

$$\begin{array}{rr} 1 & \{16\} \\ - & 8 \\ \hline & ? \end{array}$$

Subtract the ones, 16 – 8 = 8.

$$\begin{array}{rr} 1 & \{16\} \\ - & 8 \\ \hline & 8 \end{array}$$

Subtract the tens, 1 – 0 = 1 ten.

$$\begin{array}{rr} 1 & \{16\} \\ - & 8 \\ \hline 1 & 8 \end{array}$$

Therefore, 26 – 8 = 18.

To check the subtraction results, add the difference to the number that was subtracted.

Does 18 + 8 = 26? Yes.

- **Example:** Subtract 10,322 – 899.

```
   10,322
 –    899
 --------
        ?
```

Borrow from the tens' column and subtract the ones.

```
   10,31{12}
 –    89  9
 -----------
          3
```

Borrow from the hundreds' column and subtract the tens.

```
   10,2{11}{12}
 –     8  9   9
 --------------
          2   3
```

Borrow from the thousands' column and subtract hundreds.

```
   9{12}{11}{12}
 –     8   9   9
 ---------------
       4   2   3
```

Subtract the thousands.

$$\begin{array}{rrrr} 9\{12\} & \{11\} & \{12\} \\ - \quad 8 & 9 & 9 \\ \hline 9 \quad 4 & 2 & 3 \end{array}$$

or 9,423

To check the subtraction results, add the difference to the number that was subtracted.

$$\begin{array}{r} 9423 \\ + \quad 899 \\ \hline 10{,}322 \end{array}$$

• Note that if a larger number is subtracted from a smaller number, a negative number results. For example, $5 - 8 = -3$ and $40 - 50 = -10$. See Section 1.11 "Addition and Subtraction of Negative and Positive Integers."

1.5 Multiplication of Whole Numbers

This section includes definitions, an explanation of multiplication, and a detailed description of the standard multiplication technique.

- Multiplication is a shortcut for addition.
- The symbols for multiplication are ×, *, •, ()().
- The numbers to be multiplied are called the *multiplicand* (the first number) and the *multiplier* (the second number).
- The answer obtained in multiplication is called the *product*.

- Multiplication is written using the following symbols:

$$6 \times 2, 6 * 2, 6 \bullet 2, (6)(2)$$

- If you add 12 twelves, the answer is 144.

$$12 + 12 + 12 + 12 + 12 + 12 + 12 + 12 + 12 + 12 + 12 + 12 = 144$$

Adding 12 twelves is the same as multiplying twelve by twelve.

$$(12)(12) = 144$$

- What is 3 times 2, (3)(2)? It is the value of 3 two times, or $3 + 3 = 6$.

Equivalently, what is 2 times 3, (2)(3)? It is the value of 2 three times or, $2 + 2 + 2 = 6$.

- What is 2 times 4 times 3?

$$(2)(4)(3) = (2)(4 \text{ three times})$$

$$= (2)(4 + 4 + 4) = (2)(12) = 24$$

Also, $(2)(12) = 2$ twelve times

$$= 2 + 2 + 2 + 2 + 2 + 2 + 2 + 2 + 2 + 2 + 2 + 2 = 24$$

Equivalently, 2 times 4 times 3 is:

$$(2)(4)(3) = (4 \text{ two times})(3) = (4 + 4)(3)$$

$$= (8)(3) = 8 \text{ three times} = 8 + 8 + 8 = 24$$

• The order in which numbers are multiplied does not affect the result, just as the order in which numbers are added does not affect the result.

• If a number is multiplied by zero, that number is multiplied zero times and equals zero.

• For any number n, the following is true: $n \times 0 = 0$
(Note that letters are often used to represent numbers.)

An Explanation of Multiplication

• Consider the following examples:

$28 \times 7 = ?$

$6{,}846 \times 412 = ?$

$10 \times 10 = ?$

$10 \times 100 = ?$

$10 \times 1{,}000 = ?$

• To multiply numbers with two or more digits, it is easier to write the numbers in a column format. Then multiply each digit in the multiplicand (top number), beginning with the ones' digit, by each digit in the multiplier (bottom number), beginning with the ones' digit.

• **Example:** Multiply $28 \times 7 = ?$

$$\begin{array}{r} 28 \\ \times\ \ 7 \\ \hline ? \end{array}$$

This equation can be rewritten as:

$$\begin{array}{r} 2\text{ tens} + 8\text{ ones} \\ \times\ 0\text{ tens} + 7\text{ ones} \\ \hline ? \end{array}$$

Multiply the 8 ones by the 7 ones to obtain 56 ones.

$$\begin{array}{r} 2\text{ tens} + 8\text{ ones} \\ \times\ 0\text{ tens} + 7\text{ ones} \\ \hline 56\text{ ones} \end{array}$$

Multiply the 2 tens (20) by the 7 ones to obtain 14 tens (140).

$$\begin{array}{r} 2\text{ tens} + 8\text{ ones} \\ \times\ 0\text{ tens} + 7\text{ ones} \\ \hline 14\text{ tens} + 56\text{ ones} \end{array}$$

$= 140 + 56 = 196$

A Detailed Description of the Standard Multiplication Technique

- To multiply numbers, multiply each digit in the multiplier with each digit in the multiplicand (carrying-over as in addition) to create partial products, then add the partial products.

- **Example:** Multiply $28 \times 7 = ?$.

$$\begin{array}{r} 28 \\ \times\ 7 \\ \hline ? \end{array}$$

Multiply $8 \times 7 = 56$, and carry over the 5 tens.

$$\begin{array}{r} \mathit{5} \\ 28 \\ \times 7 \\ \hline 6 \end{array}$$

Multiply $2 \times 7 = 14$ tens, and add the 5 tens resulting in 19 tens.

$$\begin{array}{r} \mathit{5} \\ 28 \\ \times 7 \\ \hline 196 \end{array}$$

- **Example:** Multiply $6{,}846 \times 412 = ?$.

$$\begin{array}{r} 6846 \\ \times 412 \\ \hline ? \end{array}$$

Because there is more than one digit in the multiplier, multiply each digit in the multiplier separately to create partial products. Then, add partial products.

Note that alignment is important! Each partial product must be aligned with the right end of the multiplier digit.

First, multiply the ones' digit in the multiplier with each digit in the multiplicand beginning with the ones' digit, $6 \times 2 = 12$, carry over the 1 ten.

$$\begin{array}{r} \mathit{1} \\ 6846 \\ \times 412 \\ \hline 2 \end{array}$$

Multiply the ones' digit in the multiplier with the tens' digit in the multiplicand, $4 \times 2 = 8$, then add the 1 ten that was carried over (there is nothing new to carry).

$$\begin{array}{r} \mathit{1} \\ 6846 \\ \times \quad 412 \\ \hline 92 \end{array}$$

Multiply the ones' digit in the multiplier with the hundreds' digit in the multiplicand, $8 \times 2 = 16$, (nothing carried over to add), carry over the 1 thousand.

$$\begin{array}{r} \mathit{1}\ \mathit{1} \\ 6846 \\ \times \quad 412 \\ \hline 692 \end{array}$$

Multiply the ones' digit in the multiplier with the thousands' digit in the multiplicand, $6 \times 2 = 12$, add the 1 thousand that was carried.

$$\begin{array}{r} \mathit{1}\ \mathit{1} \\ 6846 \\ \times \quad 412 \\ \hline 13692 \end{array}$$

Resulting in the partial product aligned with the right end of the ones' digit of the multiplier.

Next, multiply the tens' digit in the multiplier with the ones' digit in the multiplicand, $6 \times 1 = 6$, (nothing to carry).

$$\begin{array}{r} 6846 \\ \times \quad 412 \\ \hline 13692 \\ 6 \end{array}$$

Multiply the tens' digit in the multiplier with the tens' digit in the multiplicand, $4 \times 1 = 4$, (nothing to add or carry).

$$\begin{array}{r} 6846 \\ \times \quad 412 \\ \hline 13692 \\ 46 \end{array}$$

Multiply the tens' digit in the multiplier with the hundreds' digit in the multiplicand, $8 \times 1 = 8$, (nothing to add or carry).

$$\begin{array}{r} 6846 \\ \times \quad 412 \\ \hline 13692 \\ 846 \end{array}$$

Multiply the tens' digit in the multiplier with the thousands' digit in the multiplicand, $6 \times 1 = 6$, (nothing carried over to add).

$$\begin{array}{r} 6846 \\ \times \quad 412 \\ \hline 13692 \\ 6846 \end{array}$$

Resulting in the partial product aligned with the right end of the tens' digit of the multiplier. (Each partial product must be aligned with the right end of the multiplier digit.)

Next, multiply the hundreds' digit in the multiplier with the ones' digit in the multiplicand, $6 \times 4 = 24$, and carry over the 2 to the tens' column.

$$
\begin{array}{r}
\textit{2} \\
6846 \\
\times \quad 412 \\
\hline
13692 \\
6846 \\
4
\end{array}
$$

Multiply the hundreds' digit in the multiplier with the tens' digit in the multiplicand, $4 \times 4 = 16$, add the 2 over the tens' column, and carry over the 1 to the hundreds' column.

$$
\begin{array}{r}
\textit{12} \\
6846 \\
\times \quad 412 \\
\hline
13692 \\
6846 \\
84
\end{array}
$$

Multiply the hundreds' digit in the multiplier with the hundreds' digit in the multiplicand, $8 \times 4 = 32$, add the 1 over the hundreds' column, and carry over the 3 to the thousands' column.

$$\begin{array}{r}
\mathit{312} \\
6846 \\
\times \quad 412 \\
\hline
13692 \\
6846 \\
384
\end{array}$$

Multiply the hundreds' digit in the multiplier with the thousands' digit in the multiplicand, $6 \times 4 = 24$, and add the 3 over the thousands' column.

$$\begin{array}{r}
\mathit{312} \\
6846 \\
\times \quad 412 \\
\hline
13692 \\
6846 \\
27384
\end{array}$$

Resulting in the partial product aligned with the right end of the hundreds' digit of the multiplier. (Each partial product is aligned with the right end of the multiplier digit.)

Next, add the three partial products for a total product.

$$\begin{array}{r}
6846 \\
\times \quad 412 \\
\hline
13692 \\
6846 \\
27384 \\
\hline
2820552
\end{array}$$

Therefore, $6{,}846 \times 412 = 2{,}820{,}552$.

- **Example:** Zeros in the multiplier.

$$\begin{array}{r} 603 \\ \times \quad 20 \\ \hline ? \end{array}$$

Multiply $3 \times 0 = 0$.

$$\begin{array}{r} 603 \\ \times \quad 20 \\ \hline 0 \end{array}$$

Multiply $0 \times 0 = 0$.

$$\begin{array}{r} 603 \\ \times \quad 20 \\ \hline 00 \end{array}$$

Multiply $6 \times 0 = 0$.

$$\begin{array}{r} 603 \\ \times \quad 20 \\ \hline 000 \end{array}$$

Resulting in the first partial product.

Multiply $3 \times 2 = 6$.

$$\begin{array}{ccccc} & & 6 & 0 & 3 \\ \times & & & 2 & 0 \\ \hline & & 0 & 0 & 0 \\ & & & 6 & \end{array}$$

Multiply $0 \times 2 = 0$.

$$\begin{array}{ccccc} & & 6 & 0 & 3 \\ \times & & & 2 & 0 \\ \hline & & 0 & 0 & 0 \\ & & 0 & 6 & \end{array}$$

Multiply $6 \times 2 = 12$.

$$\begin{array}{ccccc} & & 6 & 0 & 3 \\ \times & & & 2 & 0 \\ \hline & & 0 & 0 & 0 \\ & 1 & 2 & 0 & 6 \end{array}$$

Resulting in the second partial product.

Next, add the partial products.

$$\begin{array}{cccccc} & & & 6 & 0 & 3 \\ \times & & & & 2 & 0 \\ \hline & & & 0 & 0 & 0 \\ & 1 & 2 & 0 & 6 & \\ \hline & 1 & 2 & 0 & 6 & 0 \end{array}$$

Therefore, $603 \times 20 = 12{,}060$.

Note: When there is a zero in the multiplier, a single zero can be inserted directly below the zero in the multiplier, instead of a row of zeros.

$$\begin{array}{r} 603 \\ \times \quad 20 \\ \hline 000 \\ 1206 \\ \hline \end{array} \quad \text{Is equivalent to:} \quad \begin{array}{r} 603 \\ \times \quad 20 \\ \hline 12060 \end{array}$$

- Examples of multiplying by numbers containing zeros:

$10 \times 10 = 100$

$10 \times 100 = 1{,}000$

$10 \times 1{,}000 = 10{,}000$

$10 \times 10{,}000 = 100{,}000$

$5 \times 10 = 50$

$5 \times 100 = 500$

$5 \times 1{,}000 = 5{,}000$

$5 \times 10{,}000 = 50{,}000$

$5 \times 100{,}000 = 500{,}000$

Where,

Multiplying by 10 inserts 1 zero

Multiplying by 100 inserts 2 zeroes

Multiplying by 1,000 inserts 3 zeros

Multiplying by 10,000 inserts 4 zeros

Etc.

1.6 Division of Whole Numbers

This section provides definitions and describes in detail the long division format.

- Division evaluates how many times one number is present in another number.
- The symbols for division are $\div$, /, $\overline{)\ }$.
- The number that gets divided is called the *dividend*.
- The number that does the dividing (or divides into the dividend) is called the *divisor*.
- The answer obtained after division is called the *quotient*.
- Division is written $6 \div 2$, $\frac{6}{2}$, (6)/(2), $2\overline{)6}$.
- Division is the inverse of multiplication.

$$2 * 3 = 6, 6 \div 2 = 3, 6 \div 3 = 2$$

- Six divides by two three times or by three two times.

$$6 \div 2 = 3, 6 \div 3 = 2$$

$$6 = 3 + 3 = 2 + 2 + 2$$

- Because $6 + 6 + 6 + 6 = 24$, there are four sixes in 24.

$$6 * 4 = 24 \text{ or } 24 \div 6 = 4$$

- For any number n, the following is true: $n \div 0 =$ Undefined. (Note that letters are often used to represent numbers.)

- For any number n, the following is true: $0 \div n = 0$

A Description of the Long Division Technique

- **Example:** $4{,}628 \div 5 = ?$.

This division problem is easier to solve by writing it in a long division format.

$$5\overline{)\,4628}$$

Divide the divisor (5) into the left digit(s) of the dividend. To do this, choose the smallest part of the dividend the divisor will divide into. Because 5 does not divide into 4, the next smallest part of the dividend is 46. Estimate how many times 5 will divide into 46. First try 9. What is 5×9? $5 \times 9 = 45$, which is 1 less than 46. Write the 9 over the right end of the number it will divide into (46), and place 45 under the 46. (See Section 1.18 on rounding, truncating, and estimating for assistance on estimating.)

$$\begin{array}{r} 9??\\ 5\overline{)\,4628}\\ 45 \end{array}$$

Subtract 46 – 45.

$$\begin{array}{r} 9??\\ 5\overline{)\,4628}\\ \underline{45}\\ 01 \end{array}$$

Bring down the next digit (2) to obtain the next number (12) to divide the 5 into.

```
    9??
 5) 4628
    45
    012
```

Estimate the most number of times 5 will divide into 12. The estimate is 2. Because, $5 \times 2 = 10$, (with a remainder of 2), 5 will divide into 12 two times. Write the 2 over the right end of the number it will divide into (12), and place 10 under 12.

```
    92?
 5) 4628
    45
    012
     10
```

Subtract 12 – 10.

```
    92?
 5) 4628
    45
    012
     10
      2
```

Bring down the next digit (8) to obtain the next number (28) to divide 5 into.

$$
\begin{array}{r}
92? \\
5\overline{)\,4628} \\
\underline{45} \\
012 \\
\underline{10} \\
28
\end{array}
$$

Estimate the most number of times 5 will divide into 28. The estimate is 5. Because $5 \times 5 = 25$, (with a remainder of 3), 5 will divide into 28 five times. Write the 5 over the right end of the number it will divide into (28), and place 25 under 28.

$$
\begin{array}{r}
925 \\
5\overline{)\,4628} \\
\underline{45} \\
012 \\
\underline{10} \\
28 \\
25
\end{array}
$$

Subtract 28 – 25.

```
    925
5) 4628
   45
   012
    10
     28
     25
      3
```

There are no more numbers to bring down, therefore the division is complete, and 3 is the final remainder.

Therefore, 4,628 $\div$ 5 = 925 plus a remainder of 3.

To check division, multiply the quotient by the divisor and add the remainder to obtain the dividend.

$$\begin{array}{r} 925 \\ \times \quad 5 \\ \hline 4625 \end{array}$$

Add the remainder.

$$\begin{array}{r} 4625 \\ + \quad 3 \\ \hline 4628 \end{array}$$

- **Example:** $160{,}476 \div 364 = ?$.

Arrange in the long division format.

$$
\begin{array}{r}
? \\
364\overline{)160476}
\end{array}
$$

Divide the divisor (364) into the left digits of the dividend. To do this, choose the smallest part of the dividend the divisor will divide into. 364 will divide into 1,604. Estimate the most number of times 364 will divide into 1,604. The estimate is 4. $364 \times 4 = 1{,}456$ with a remainder of 148. Write the 4 over the right end of the number it will divide into (1,604) and place 1,456 under the 1,604.

$$
\begin{array}{r}
4?? \\
364\overline{)160476} \\
1456
\end{array}
$$

Subtract 1,604 – 1,456.

$$
\begin{array}{r}
4?? \\
364\overline{)160476} \\
\underline{1456} \\
0148
\end{array}
$$

Bring down the next digit (7) to obtain the next number (1,487) to divide 364 into.

$$
\begin{array}{r}
4?? \\
364\overline{)160476} \\
\underline{1456} \\
01487
\end{array}
$$

Estimate the most number of times 364 will divide into 1,487. The estimate is 4. $364 \times 4 = 1{,}456$ with a remainder of 31. Write the 4 over the end of the number it will divide into (1,487), and place 1,456 under the 1,487.

$$
\begin{array}{r}
44? \\
364\overline{)160476} \\
\underline{1456} \\
01487 \\
\underline{1456}
\end{array}
$$

Subtract 1,487 – 1,456.

$$
\begin{array}{r}
44? \\
364\overline{)160476} \\
\underline{1456} \\
01487 \\
\underline{1456} \\
31
\end{array}
$$

Bring down the next digit (6) to obtain the next number (316) to divide 364 into.

$$
\begin{array}{r}
44? \\
364\overline{)160476} \\
\underline{1456} \\
01487 \\
\underline{1456} \\
316
\end{array}
$$

Because 364 will not divide into 316, place a zero over the right end of 316.

$$
\begin{array}{r}
440 \\
364 \overline{)\,160476} \\
\underline{1456} \\
01487 \\
\underline{1456} \\
316
\end{array}
$$

316 becomes the remainder. Therefore, 160,476 ÷ 364 = 440 plus a remainder of 316.

To check division, multiply the quotient by the divisor and add the remainder to obtain the dividend.

$$
\begin{array}{r}
364 \\
\times \quad 440 \\
\hline
14560 \\
\underline{1456} \\
160160
\end{array}
\qquad \text{Add the remainder.} \qquad
\begin{array}{r}
160160 \\
+ \quad 316 \\
\hline
160476
\end{array}
$$

1.7 Divisibility, Remainders, Factors, and Multiples

This section defines and gives examples of divisibility, remainders, factors, and multiples.

- The *divisibility* of a number is determined by how many times that number can be divided evenly by another number. For example, 6 is divisible by 2 because 2 divides into 6 a total of 3 times with no remainder. An example of a number that is not divisible by 2 is 5, because 2 divides into 5 a total of 2 times with 1 remaining.

- The *remainder* is the number left over when a number cannot be divided evenly by another number.

- A smaller number is a *factor* of a larger number if the smaller number can be divided into the larger number without producing a remainder. Numbers that are multiplied together to produce a product are factors of that product.

- The factors of 6 are 6 and 1, or 2 and 3.

$$6 \div 6 = 1, \quad 6 \div 1 = 6, \quad 6 \div 2 = 3, \quad 6 \div 3 = 2.$$

$$6 \times 1 = 6, \quad 2 \times 3 = 6$$

5 is not a factor of 6 because $6 \div 5 = 1$ with a remainder of 1.

- The factors of 10 are 2 and 5, or 10 and 1, where:

$$10 = (2)(5), \quad 10 = (10)(1)$$

- If a and b represent numbers,

$$6ab = (2)(3)(a)(b)$$

Where 2, 3, a, and b are factors of 6ab.

- A *multiple* of a number is any number that results after that number is multiplied with any number. Multiples of zero do not exist because any number multiplied by zero, including zero, results in zero. It is possible to create infinite multiples of numbers by simply multiplying them with other numbers.

- The following are examples of multiples of the number 3:

$$3 \times 5 = 15$$

$$3 \times 2 = 6$$

$$3 \times 4 = 12$$

$$3 \times 20 = 60$$

Where 15, 6, 12 and 60 are some of the multiples of 3.

1.8 Integers

- Integers include positive numbers and zero (whole numbers) and also negative numbers.

- The set of all integers is represented as follows:

$$\text{Integers} = \{\ldots -6, -5, -4, -3, -2, -1, 0, 1, 2, 3, 4, 5, 6, 7, \ldots\}$$

• Numbers that are *not* integers include numbers in the form of fractions, decimals, percents, or exponents.

• Consecutive integers are integers that are arranged in an increasing order according to their size from the smallest to the largest without any integers missing in between.

• The following are examples of consecutive integers:

$\{-10, -9, -8, -7\}$

$\{0, 1, 2, 3, 4\}$

$\{-2, -1, 0, 1, 2, 3, 4, 5\}$

$\{99, 100, 101, 102, 103, 104, 105\}$

1.9 Even and Odd Integers

This section defines and provides examples of even integers and odd integers.

• Even integers are integers that can be divided evenly by 2.

Even Integers $= \{... -6, -4, -2, 0, 2, 4, 6, 8, ...\}$

• Zero is an even integer.

• Odd integers are integers that cannot be divided evenly by 2, and therefore are not even.

Odd Integers $= \{... -7, -5, -3, -1, 1, 3, 5, 7, ...\}$

- Fractions are neither even nor odd. If division of two numbers yields a fraction, the result is not odd or even.

- It is possible to immediately determine if a large number is even or odd by observing whether the digit in the ones position is even or odd.

- Consecutive even integers are even integers that are arranged in an increasing order according to their size without any integers missing in between.

  The following is an example of consecutive even integers:
  $\{-10, -8, -6, -4, -2, 0, 2, 4, 6\}$

- Consecutive odd integers are odd integers that are arranged in an increasing order according to their size without any integers missing in between.

  The following is an example of consecutive odd integers:
  $\{-3, -1, 1, 3, 5\}$

- Adding and subtracting even and odd integers:

  even + even = even, $4 + 6 = 10$, $0 + 6 = 6$

  even + odd = odd, $4 + 5 = 9$, $0 + 5 = 5$

  odd + odd = even, $3 + 5 = 8$, $1 + 5 = 6$

  even – even = even, $6 - 4 = 2$, $4 - 6 = -2$

  even – odd = odd, $6 - 3 = 3$, $4 - 3 = 1$

  odd – odd = even, $5 - 3 = 2$, $1 - 3 = -2$

- Multiplying even and odd integers:

$$\text{even} \times \text{even} = \text{even}, \quad 4 \times 6 = 24, \quad 0 \times 6 = 0$$

$$\text{even} \times \text{odd} = \text{even}, \quad 4 \times 5 = 20, \quad 0 \times 5 = 0$$

$$\text{odd} \times \text{odd} = \text{odd}, \quad 5 \times 5 = 25, \quad 1 \times 5 = 5$$

1.10 Zero

- Zero is both an integer and a whole number.
- Zero is even.
- Zero is not a positive or negative number.
- If zero is added to or subtracted from any number, the value of that number is not changed and the result is that number:

$$n + 0 = n \text{ and } n - 0 = n$$

- If zero is multiplied by any number the result is zero:

$$n \times 0 = 0$$

- Dividing by zero is "undefined":

$$n \div 0 = \text{undefined}$$

- Zero divided by a number (represented by letter n) is zero:

$$0 \div n = 0$$

1.11 Addition and Subtraction of Negative and Positive Integers

This section describes addition and subtraction of negative integers and positive integers.

- Remember the number line.

$$-5 \quad -4 \quad -3 \quad -2 \quad -1 \quad 0 \quad +1 \quad +2 \quad +3 \quad +4 \quad +5$$

- When numbers are added and subtracted, think of moving along the number line.

- Begin with the first number and move to the right or left depending on the sign of the second number and whether it is being added or subtracted to the first number.

- To add a positive number, begin at the first number and move to the right the value of the second number.

$2 + 1 = 3$

$$-5 \quad -4 \quad -3 \quad -2 \quad -1 \quad 0 \quad +1 \quad +2 \quad +3 \quad +4 \quad +5$$

$-2 + 1 = -1$

$$-5 \quad -4 \quad -3 \quad -2 \quad -1 \quad 0 \quad +1 \quad +2 \quad +3 \quad +4 \quad +5$$

• To add a negative number (subtract), begin at the first number and move to the left the value of the second number.

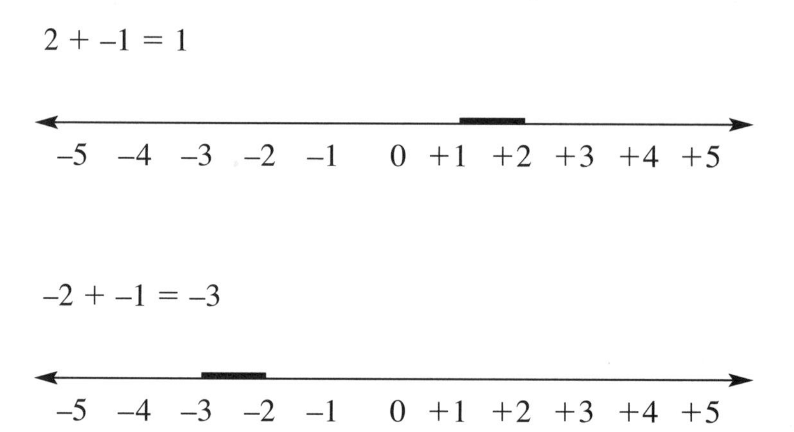

• To subtract a positive number, begin at the first number and move to the left the value of the second number.

$2 - 1 = 1$

–5 –4 –3 –2 –1 0 +1 +2 +3 +4 +5

$-2 - 1 = -3$

–5 –4 –3 –2 –1 0 +1 +2 +3 +4 +5

• To subtract a negative number, begin at the first number and move to the right the value of the second number, (the two negatives cancel each other out).

$2 - -1 = 2 + 1 = 3$

–5 –4 –3 –2 –1 0 +1 +2 +3 +4 +5

$-2 - -1 = -2 + 1 = -1$

–5 –4 –3 –2 –1 0 +1 +2 +3 +4 +5

• If a and b represent numbers,

$a + +b = a + b$

$a + -b = a - b$

$a - +b = a - b$

$a - -b = a + b$

1.12 Multiplication and Division of Negative and Positive Integers

• If a negative number and a positive number are multiplied or divided, the result will be negative. If two negative numbers are multiplied or divided, the negative signs will cancel each other and the resulting number will be positive.

- The following rules apply to multiplication and division of positive and negative numbers, (a and b represent numbers):

positive × positive = positive, $3 \times 2 = 6$, $a \times b = ab$

negative × negative = positive, $-3 \times -2 = 6$, $-a \times -b = ab$

positive × negative = negative, $3 \times -2 = -6$, $a \times -b = -ab$

negative × positive = negative, $-3 \times 2 = -6$, $-a \times b = -ab$

positive ÷ positive = positive, $^6/_2 = 3$, $^a/_b = {}^a/_b$

negative ÷ negative = positive, $^{-6}/_{-2} = 3$, $^{-a}/_{-b} = {}^a/_b$

positive ÷ negative = negative, $^6/_{-2} = -3$, $^a/_{-b} = {}^{-a}/_b$

negative ÷ positive = negative, $^{-6}/_2 = -3$, $^{-a}/_b = {}^{-a}/_b$

- A summary of the multiplication rules are:

(+)(+) = +

(–)(–) = +

(+)(–) = –

(–)(+) = –

- A summary of the division rules are:

(+) ÷ (+) = +

(–) ÷ (–) = +

(+) ÷ (–) = –

(–) ÷ (+) = –

1.13 The Real Number Line

- Real numbers include whole numbers, integers, fractions, decimals, rational numbers, and irrational numbers.
- Real numbers can be expressed as the sum of a decimal and an integer.
- All real numbers except zero are either positive or negative.
- All real numbers correspond to points on the real number line, and all points on the number line correspond to real numbers.
- The real number line reaches from negative infinity ($-\infty$) to positive infinity ($+\infty$).

$-4 \quad -3 \quad -2 \quad -1 \; -.5 \quad 0 \quad 1 \; \sqrt{2} \; 2 \; 5/2 \; 3 \; \pi \quad 4$

Real numbers include -0.5, $\sqrt{2}$, $5/2$ and π. ($\pi = 3.14$)
All numbers to the left of zero are negative.
All numbers to the right of zero are positive.

- The distance between zero and a number on the number line is called the absolute value or the magnitude of the number.

1.14 Absolute Value

- The absolute value is the distance between zero and the number on the number line.

$-4 \quad -3 \quad -2 \quad -1 \; -.5 \quad 0 \quad 1 \; \sqrt{2} \; 2 \; 5/2 \; 3 \; \pi \quad 4$

- The absolute value is always positive or zero, never negative.
- The symbol for absolute value of a number represented by n is $|n|$.
- Positive 4 and negative 4 have the same absolute value.

$|4| = 4$ and $|-4| = 4$

- Properties of absolute value are (x and y represent numbers):

$|x| \geq 0$

$|x - y| = |y - x|$

$|x||y| = |xy|$

$|x + y| \leq |x| + |y|$

(See Section 1.19 for description of $\leq$ and $\geq$.)

1.15 Prime Numbers

- A prime number is a number that can only be divided evenly (not producing a remainder) by itself and by 1.
- For example, 7 can only be divided evenly by 7 and by 1.
- Examples of prime numbers are:

$\{2, 3, 5, 7, 11, 13, 17, 19, 23\}$

- Zero and 1 are *not* prime numbers.
- The only even prime number is 2.

1.16 Rational vs. Irrational Numbers

In this section, rational numbers and irrational numbers are defined.

• A number is a *rational number* if it can be expressed in the form of a fraction, x/y, and the denominator is not zero.

• For example, 2 is a rational number because it can be expressed as the fraction 2/1.

• Every integer can be expressed as a fraction and is a rational number.

(Integer)/1 or n/1 Where n = any integer.

• Whole numbers are included in the set of integers, and whole numbers are rational numbers.

• A number is an *irrational number* if it is not a rational number and therefore cannot be expressed in the form of a fraction.

• Examples of irrational numbers are numbers that possess endless non-repeating digits to the right of the decimal point, such as,

$\pi = 3.1415...$, and $\sqrt{2} = 1.414...$

1.17 Complex Numbers

In this section, complex numbers, real numbers and imaginary numbers are defined, and addition, subtraction, multiplication and division of complex numbers are described.

• *Complex numbers* are numbers involving $\sqrt{-1}$, (see chapter on roots and radicals for a description of $\sqrt{\ }$). There is no number that when squared equals –1. By definition $(\sqrt{-1})^2$ should equal –1, but it does not, therefore the symbol i is introduced, such that:

$\sqrt{-x} = i\sqrt{x}$ Where x is a positive number and $(i)^2 = -1$.

• For example, evaluate $(\sqrt{-5})^2$.

$$(\sqrt{-5})^2 = (i\sqrt{5})(i\sqrt{5}) = (i)^2(\sqrt{5})(\sqrt{5})$$

$$= (i)^2\sqrt{(5)(5)} = (-1)(5) = -5$$

• Complex numbers are numbers involving i and are generally in the form:

$x + iy$ Where x and y are *real numbers*.

• In the expression, $x + iy$, x is called the real part and iy is called the imaginary part.

• A real number multiplied by i forms an *imaginary number.*

(real number)(i) = imaginary number

• A real number added to an imaginary number forms a *complex number.*

(real number) + (real number)(i) = complex number

• To add or subtract complex numbers, add or subtract the real parts and the imaginary parts separately.

- **Example:** Add $(5 + 4i) + (3 + 2i)$.

$$(5 + 4i) + (3 + 2i) = (5 + 3) + (4i + 2i) = 8 + 6i$$

- **Example:** Subtract $(5 + 4i) - (3 + 2i)$.

$$(5 + 4i) - (3 + 2i) = (5 - 3) + (4i - 2i) = 2 + 2i$$

- Complex numbers are multiplied as ordinary binomials, and $(i)^2$ is replaced by -1. (See Chapter 5, "Polynomials," Section 5.4 for multiplication of polynomials in the second book of the *Master Math* series entitled *Algebra*.) To multiply binomials, each term in the first binomial is multiplied by each term in the second binomial, and like terms are combined (added).

- **Example:** Multiply $(5 + 4i) \times (3 + 2i)$.

$$(5 + 4i)(3 + 2i)$$

$$= (5)(3) + (5)(2i) + (4i)(3) + (4i)(2i)$$

$$= 15 + 10i + 12i + 8(i)^2 = 15 + 22i + 8(-1)$$

$$= 15 + 22i - 8 = (15 - 8) + 22i = 7 + 22i$$

- To divide complex numbers, first multiply the numerator and denominator by what is called the *complex conjugate* of the denominator. The complex conjugate of $(3 + 2i)$ is $(3 - 2i)$, and the complex conjugate of $(3 - 2i)$ is $(3 + 2i)$. The product of a complex number and its conjugate is a real number. Remember to replace $(i)^2$ by -1. (See chapters 5 entitled "Polynomials" and 6 entitled "Algebraic Fractions with Polynomial Fractions" for division and multiplication of polynomials and polynomial fractions in the second book of the *Master Math* series entitled *Algebra*.)

• **Example:** Divide (5 + 4i) ÷ (3 + 2i), (note that (3 – 2i) is the complex conjugate of the denominator).

$$\frac{(5+4i)}{(3+2i)} = \frac{(5+4i)(3-2i)}{(3+2i)(3-2i)}$$

$$= \frac{(5)(3)+(5)(-2i)+(4i)(3)+(4i)(-2i)}{(3)(3)+(3)(-2i)+(2i)(3)+(2i)(-2i)}$$

$$= \frac{15+-10i+12i+-8i^2}{9+-6i+6i+-4i^2}$$

$$= \frac{(15+-8(i)^2)+(-10i+12i)}{(9+-4(i)^2)+(-6i+6i)}$$

$$= \frac{(15+-8(-1))+2i}{(9+-4(-1))+0} = \frac{15+8+2i}{9+4+0}$$

$$= \frac{23+2i}{13} = \frac{23}{13} + \frac{2i}{13}$$

Note that the answer is not in the form (23 + 2i)/(13) because this is a complex number.

1.18 Rounding, Truncating, and Estimating Numbers

• Quickly estimating the answer to a problem can be achieved by rounding the numbers that are to be added, subtracted, multiplied, or divided, then adding, subtracting, multiplying, or dividing the rounded numbers. Numbers can be rounded to the nearest ten, hundred, thousand, million, etc., depending on their size.

• For example, estimate the sum of $(38 + 43)$ by rounding.

Round 38 to the nearest ten. 40

Round 43 to the nearest ten. 40

The estimated sum is $40 + 40 = 80$

Compare the estimate with the actual sum.

$38 + 43 = 81$

80 is a good estimate of the actual value, 81.

• To round a number to the nearest ten, hundred, thousand, etc., the last retained digit should either be increased by one or left unchanged according to the following rules:

If the left most digit to be dropped is less than 5, leave the last retained digit unchanged.

If the left most digit to be dropped is greater than 5, increase the last retained digit by one.

If the left most digit to be dropped is equal to 5, leave the last retained digit unchanged if it is even or increase the last retained digit by one if it is odd.

• **Example:** Round the following numbers to the nearest ten.

11 rounds to 10

65 rounds to 60

538 rounds to 540

65,236 rounds to 65,240

• **Example:** Round the following numbers to the nearest hundred.

238 rounds to 200

650 rounds to 600

9,436 rounds to 9,400

750 rounds to 800

740 rounds to 700

• **Example:** Round the following numbers to the nearest tenth. (See Chapter 3, "Decimals.")

5.01 rounds to 5.0

8.38 rounds to 8.4

9.55 rounds to 9.6

9.65 rounds to 9.6

- **Example:** Estimate the difference of (956 – 838) by rounding to the nearest hundred and to the nearest ten.

Round to the nearest hundred.

956 rounds to 1,000

838 rounds to 800

Subtract rounded numbers. 1,000 – 800 = 200

Round to the nearest ten.

956 rounds to 960

838 rounds to 840

Subtract rounded numbers. 960 – 840 = 120

Compare with the actual values. 956 – 838 = 118

Rounding to the nearest hundred did not produce as good of an estimate as rounding to the nearest ten.

- **Example:** Estimate the product of (54×2) by rounding.

Round 54 to 50

Multiply rounded numbers. $50 \times 2 = 100$

Compare with the actual numbers. $54 \times 2 = 108$

- **Example:** Estimate the quotient of $(43 \div 2)$ by rounding.

  Round 43 to 40

  Divide rounded numbers. $40 \div 2 = 20$

  Compare with the actual numbers. $43 \div 2 = 21.5$

- **Example:** Estimate the sum of $(56 + 68 + 43)$ by rounding

  Round each number and add the estimate.

  $60 + 70 + 40 = 170$

  Compare with the actual numbers.

  $56 + 68 + 43 = 167$

- Truncating rounds down to the smaller ten, hundred, thousand, etc. For example,

  43 truncates to 40

  768 truncates to 760

1.19 Inequalities, $>$, $<$, $\geq$, $\leq$

- Inequalities are represented by the symbols for greater than and less than, and describe expressions in which the value on one side of the symbol is greater than the value on the other side of the symbol.

- The symbol for greater than is $>$
- The symbol for less than is $<$
- The symbol for greater than or equal to is $\geq$
- The symbol for less than or equal to is $\leq$

- If a and b are numbers,

  $a < b$ represents a is less than b, or a is to the left of b on the number line.

  $a > b$ represents a is greater than b, or a is to the right of b on the number line.

  $a \leq b$ represents a is less than or equal to b.

  $a \geq b$ represents a is greater than or equal to b.

  $a < c < b$ represents a is less than c and c is less than b.

  $a > c > b$ represents a is greater than c and c is greater than b.

- **Examples:**

  $2 < 4$

  $4 > 2$

  $5 < 8$

  $8 > 5$

  $2 < 5 < 8$

  $5 \leq 8$

  $5 \leq 5$

- The number line below describes $x > 1$.

–5 –4 –3 –2 –1 0 +1 +2 +3 +4 +5

- If c is a positive number, then $c > 0$. For example, if $c = 5$, then $5 > 0$.
- If d is a negative number, then $d < 0$. For example, if $d = -5$, then $-5 < 0$.

- **Example:** If a, b, and c represent numbers, and

  If $a < b$

  Then $a + c < b + c$

  Also $a - c < b - c$

- **Example:** If $a = 2$, $b = 3$, and $c = 4$, and

  Because $2 < 3$

  Then $2 + 4 < 3 + 4$

  Equivalently $6 < 7$

  Also $2 - 4 < 3 - 4$

  Equivalently $-2 < -1$

  (Remember -2 is to the left of -1 on the number line.)

- **Example:** If a, b, and c represent numbers, and

  If $a > b$

  And if $c > 0$

  Then $a + c > b + c$

  Also $a - c > b - c$

If $a < b$

And if $c > 0$

Then $ac < bc$

If $a > b$

And if $c > 0$

Then $ac > bc$

- **Example:** If $a = 2$, $b = 3$, and $c = 4$, and

  Because $2 < 3$

  Then $(2)(4) < (3)(4)$

  Equivalently $8 < 12$

- Multiplying or dividing by a negative number causes the inequality sign to reverse.

- **Example:** If a, b, and d represent numbers, and

  If $d < 0$

  And if $a < b$

  Then $ad > bd$

  (If a negative number is multiplied, the inequality reverses.)

- **Example:** If $a = 2$, $b = 3$, and $d = -4$, and

  Because $-4 < 0$

  And $2 < 3$

  Then $(2)(-4) > (3)(-4)$

  Equivalently $-8 > -12$

  (Remember, –12 is to the left of –8 on the number line.)

- **Example:** If a, b, and d represent numbers, and

  If $d < 0$

  And if $a > b$

  Then $a \div d < b \div d$

  (If a negative number is divided, the inequality reverses.)

- **Example:** If $a = 4$, $b = 6$, and $d = -2$, and

  Because $-2 < 0$

  And $4 < 6$

  Then $4 \div -2 > 6 \div -2$

  Equivalently $-2 > -3$

  (Remember, –2 is to the right of –3 on the number line.)

- In summary, for an inequality:

If a number is added to both sides, the inequality remains unchanged.

If a number is subtracted from both sides, the inequality remains unchanged.

If a positive number is multiplied, the inequality sign remains unchanged.

If a positive number is divided, the inequality sign remains unchanged.

If a negative number is multiplied, the inequality sign reverses.

If a negative number is divided, the inequality sign reverses.

1.20 Factorial

- The factorial of a positive integer is the product of that integer and each consecutive positive integer from one to that number.
- The symbol for factorial is "!" (the explanation point).
- Examples of the factorial of numbers are:

  $5! = 1 \times 2 \times 3 \times 4 \times 5 = 120$

  $7! = 1 \times 2 \times 3 \times 4 \times 5 \times 6 \times 7 = 5{,}040$

  $3! = 1 \times 2 \times 3 = 6$

  $n! = 1 \times 2 \times 3 \times \ldots n$

  $0! = 1$ by definition.

Notes

Chapter 2

Fractions

2.1 Definitions

- A fraction is a number that is expressed in the form a/b or $\frac{a}{b}$.

- A fraction can also be defined as:

$$\frac{\text{part}}{\text{whole}}, \frac{\text{section}}{\text{whole}} \text{ or } \frac{\text{some}}{\text{all}}$$

- If a pie is cut into 8 pieces, then 2 pieces are a fraction of the whole pie.

$$\frac{\text{2 pieces}}{\text{whole pie}} = \frac{\text{2 pieces}}{\text{8 pieces in whole pie}}$$

- The top number is the *numerator*, and the bottom number is the *denominator*.

$$\frac{\text{Numerator}}{\text{Denominator}} = \text{Numerator} \div \text{Denominator}$$

- Fractions express division. $\frac{5}{8} = 5 \div 8$

- Equivalent fractions are fractions that have equal value. The following are equivalent fractions:

1/2 = 2/4 = 4/8 = 8/16 = 16/32 = 32/64

2/3 = 4/6 = 8/12 = 16/24 = 32/48

3/5 = 6/10 = 12/20 = 24/40 = 48/80

2.2 Multiplying Fractions

• To multiply fractions, multiply the numerators, multiply the denominators, and place the product of the numerators over the product of the denominators.

• **Example:**

$$\frac{9}{2} \times \frac{5}{8} \times \frac{3}{4} = \frac{9 \times 5 \times 3}{2 \times 8 \times 4} = \frac{135}{64}$$

$$\frac{2}{3} \times \frac{3}{4} = \frac{2 \times 3}{3 \times 4} = \frac{6}{12} = \frac{1}{2}$$

• If the numerators are less than the denominators, then after multiplying, the value of the product will be less than the values of the original fractions. This is true because multiplying a number or fraction by a fraction is equivalent to taking a fraction of the first number or fraction.

• For example: If you have a piece of a pie (which means you have a fraction of a pie), then you eat a fraction of the piece, you have eaten less than a piece. In other words, you have eaten a fraction of a fraction.

2.3 Adding and Subtracting Fractions with Common Denominators

• To add or subtract two or more fractions that have the same denominators, simply add or subtract the numerators and place the sum or difference over the *common denominator.*

- **Examples:**

$$\frac{9}{8} + \frac{5}{8} + \frac{3}{8} = \frac{9+5+3}{8} = \frac{17}{8}$$

$$\frac{9}{8} - \frac{5}{8} - \frac{3}{8} = \frac{9-5-3}{8} = \frac{1}{8}$$

2.4 Adding and Subtracting Fractions with Different Denominators

- To add or subtract two or more fractions that have different denominators, a common denominator must be found. Then, each original fraction must be multiplied by multiplying-fractions that have their numerator equal to their denominator, so that new equivalent fractions are created that have the same common denominator. These fractions can then be added or subtracted as shown above. The procedure is:

1. Find a common denominator for two or more fractions by calculating multiples of each denominator until one number is obtained that is a multiple of each denominator. This is called a *common multiple*. For example, if there are two denominators, 4 and 6, the multiples of 4 are: 4, 8, 12, 16, ..., and the multiples of 6 are: 6, 12, 18, 24, ... The smallest number that is a multiple of both 4 and 6 is 12. Therefore, 12 is called the lowest common multiple and it is also the *lowest common denominator* for 4 and 6.

2. After the lowest common denominator is found, transform the original fractions into fractions with common denominators so that they can be added or subtracted. To do this, multiply each original fraction by a *multiplying-fraction* that has its numerator equal to its denominator, so as to create *new equivalent fractions* that have the same common denominators. By having the numerator equal to the denominator in these multiplying-fractions, the value of each of the original fractions remains unchanged. For example, $(1/2) \times (2/2) = (2/4)$, where 1/2 is the original fraction, 2/2 is the multiplying-fraction, and 2/4 is the resulting equivalent fraction with its value equal to 1/2.

   To determine each multiplying-fraction, compare the new common denominator with the original denominators of each fraction, then create multiplying-fractions for each original fraction such that when the original denominator is multiplied with the multiplying-fraction's denominator, the resulting denominator is the common denominator.

3. After the equivalent fractions with their common denominators have been obtained, add or subtract the numerators and place the sum or difference over the common denominator.

4. Reduce the resulting fraction if it is possible, (see section below on reducing fractions).

- **Example:** Add $1/4 + 1/6 = ?$.

First, find the lowest common denominator.

Multiples of 4: 4, 8, 12, 16, ...

Multiples of 6: 6, 12, 18, ...

The lowest common multiple and lowest common denominator is 12.

Next, multiply each original fraction by a multiplying fraction (with its numerator equal to its denominator) such that each equivalent fraction will have a denominator of 12.

For the first fraction, because $4 \times 3 = 12$,

$1/4 + ?/? = ?/12$

$1/4 \times 3/3 = 3/12$

3/12 is equivalent to 1/4.

For the second fraction, because $6 \times 2 = 12$,

$1/6 + ?/? = ?/12$

$1/6 \times 2/2 = 2/12$

2/12 is equivalent to 1/6.

After the equivalent fractions with their common denominators have been obtained, add the numerators and place the sum over the common denominator.

$3/12 + 2/12 = 5/12$

- **Example:** Add

$$\frac{9}{2} + \frac{5}{5} + \frac{3}{10} = \frac{?}{?}$$

First, find the lowest common denominator.

Multiples of 2: 2, 4, 6, 8, 10, ...

Multiples of 5: 5, 10, 15, 20, ...

Multiples of 10: 10, 20, ...

The lowest common multiple is 10, therefore the lowest common denominator is 10. (Notice that the smallest number that 2, 5, and 10 will all divide into is 10.) (Also, the factors of 10 are: (2)(5) and (1)(10).)

Next, multiply each original fraction by a multiplying-fraction (with its numerator equal to its denominator) such that each equivalent fraction will have a denominator of 10.

For the first fraction, because $2 \times 5 = 10$:

$$\frac{9}{2} \times \frac{5}{5} = \frac{45}{10}$$

For the second fraction, because $5 \times 2 = 10$:

$$\frac{5}{5} \times \frac{2}{2} = \frac{10}{10}$$

For the third fraction, because $10 \times 1 = 10$:

$$\frac{3}{10} \times \frac{1}{1} = \frac{3}{10}$$

Next, add the new equivalent fractions with their common denominators (add the numerators and place the sum over the common denominator).

$$\frac{45}{10} + \frac{10}{10} + \frac{3}{10} = \frac{45 + 10 + 3}{10} = \frac{58}{10} = 29/5$$

- **Example:** Subtract 9/2 – 5/8 – 3/4.

$$\frac{9}{2} - \frac{5}{8} - \frac{3}{4} = \frac{?}{?}$$

Find the lowest common denominator.

Multiples of 2: 2, 4, 6, 8, 10, ...

Multiples of 8: 8, 16, ...

Multiples of 4: 4, 8, 12, ...

The lowest common multiple is 8, therefore the lowest common denominator is 8.

To find the multiply-fractions, consider the denominators:

$2 \times 4 = 8$

$8 \times 1 = 8$

$4 \times 2 = 8$

The multiplying-fractions are:

$$\frac{4}{4}, \frac{1}{1}, \text{ and } \frac{2}{2}$$

Therefore, the equation is:

$$\frac{9 \times 4}{2 \times 4} - \frac{5 \times 1}{8 \times 1} - \frac{3 \times 2}{4 \times 2} =$$

$$\frac{36}{8} - \frac{5}{8} - \frac{6}{8} = \frac{36 - 5 - 6}{8} = \frac{25}{8}$$

- An alternative method for adding and subtracting two fractions is as follows:

1. Multiply the two denominators to find a common denominator. The result will be a common multiple of each, but not necessarily the lowest common multiple.

   Find the new numerator of the first fraction by multiplying the numerator of the first fraction with the original denominator of the second fraction.

   Find the new numerator of the second fraction by multiplying the numerator of the second fraction with the original denominator of the first fraction.

2. Then, add or subtract the new numerators and place the result over the common denominator.
3. Reduce the resulting fraction if it is possible (see section below on reducing fractions).

• **Example:** Add 1/2 + 3/4.

Multiply denominators. $2 \times 4 = 8$

The new common denominator is 8.

The first new numerator is $1 \times 4 = 4$.

The second new numerator is $3 \times 2 = 6$.

The new equivalent fractions are:

$$\frac{4}{8} + \frac{6}{8} = \frac{10}{8}$$

To reduce 10/8, divide 2 into both the numerator and the denominator.

$$\frac{10/2}{8/2} = \frac{5}{4}$$

2.5 Dividing Fractions

• To divide fractions, turn the second fraction upside down to invert the numerator and denominator. (An inverted fraction is called a reciprocal.) Then, multiply the first fraction with the reciprocal fraction.

$$\frac{3}{4} \div \frac{5}{8} = \frac{3}{4} \times \frac{8}{5} = \frac{24}{20} = \frac{6}{5}$$

Where 8/5 is the reciprocal of 5/8.

- Multiplying the reciprocal is equivalent to dividing because:

$$3/4 \div 5/8 = \frac{3/4}{5/8} = \frac{3}{4 \times 5/8} = \frac{3 \times 8}{4 \times 5}$$

$$\frac{a}{b/c} = \frac{a \times c}{b}$$ a, b, and c represent numbers.

$$\frac{a/d}{b/c} = \frac{a \times c}{b \times d}$$ a, b, c, and d represent numbers.

2.6 Reducing Fractions

- Reducing fractions is an important step in the process of problem-solving, and can be performed before, during or after the primary operations are completed. After fractions have been added, subtracted, multiplied, or divided, the resulting fractions may be large. Large fractions can often be reduced by dividing-out or canceling factors that are contained in both the numerator and denominator.

- **Example:** Reduce 25/10.

  Factor 25 and 10.

  $25 = 5 \times 5$

  $10 = 5 \times 2$

There is a factor of 5 contained in both the numerator and denominator that can be canceled.

$$\frac{25}{10} = \frac{5 \times 5}{2 \times 5} = \frac{5}{2}$$

- **Example:** Reduce 30/40.

Write the fraction in factored form and cancel the factors common to both the numerator and denominator.

$$\frac{30}{40} = \frac{3 \times 2 \times 5}{2 \times 2 \times 2 \times 5} = \frac{3}{4}$$

Note that by inspection a 10 could have been canceled from both the numerator and the denominator.

2.7 Complex Fractions, Mixed Numbers, and Improper Fractions

In this section, improper fractions, mixed numbers, and complex fractions are defined, and methods for converting improper fractions to mixed numbers and converting mixed numbers to improper fractions are described.

- Every integer can be expressed as a fraction. For example, $6 = 6/1, 23 = 23/1$.

- *Improper fractions* are fractions with their numerators larger than their denominators. Examples of improper fractions are:

$$\frac{13}{2}, \frac{6}{1}, \frac{12}{2}$$

• If a value is represented by an integer and a fraction, it is called a *mixed number*. A mixed number always has an integer and a fraction. The following are examples of mixed numbers:

$$6\frac{1}{2},\ 25\frac{3}{4}$$

• An improper fraction can be expressed as a mixed number by dividing the numerator with the denominator.

• **Example:**

$$\frac{4}{3} = 1\frac{1}{3}$$

Where $\frac{4}{3}$ is the improper fraction and $1\frac{1}{3}$ is the mixed number.

• When performing calculations, it is generally easier to work with improper fractions than with mixed numbers.

• Mixed numbers are easily converted to improper fractions. Following are two methods used to convert mixed numbers into improper fractions.

Method 1

1. Multiply the integer and the denominator.
2. Add the numerator to the result, which results in a new numerator.
3. Place the new numerator over the original denominator.

- **Example:** Convert 6 1/2 to an improper fraction.

  Multiply the integer and the denominator. $(6)(2) = 12$

  Add the numerator to the result to obtain the new numerator. $1 + 12 = 13$

  Place the new numerator over the denominator to obtain the improper fraction. 13/2

Method 2

1. Find the common denominator.
2. Add the fractions.

- **Example:** Convert 6 1/2 to an improper fraction.

$$6\frac{1}{2} = 6 + \frac{1}{2} = \frac{6}{1} + \frac{1}{2}$$

The lowest common denominator is 2. Create equivalent fractions with common denominators of 2 and add. (1/2 does not need to be multiplied.)

$$\frac{6}{1} \times \frac{2}{2} + \frac{1}{2} = \frac{12}{2} + \frac{1}{2} = \frac{13}{2}$$

Therefore, 6 1/2 = 13/2.

- Improper fractions are easily converted to mixed numbers by dividing the numerator with the denominator.

- **Example:** Convert 25/4 to a mixed fraction.

  Divide the numerator by the denominator. $25 \div 4 = 6$ plus a remainder of 1.

  Place the remainder over the divisor resulting in 6 1/4.

  To check the result, multiply the integer and the denominator. $6 \times 4 = 24$

  Add the numerator and place the result over the denominator. $24 + 1 = 25$

  25/4 is the original improper fraction.

- A *complex fraction* has a fraction in the numerator or in the denominator or in both so that there is one or more fractions within a fraction.

- The following are examples of complex fractions (x and y represent numbers):

$$\frac{1/3}{4}, \frac{3}{4/5}, \frac{x + 2/y}{5}, \frac{3}{x/y}$$

- Complex fractions can be simplified by performing the indicated division of the fractions or sub-fractions. For example, simplify the following (remember when dividing fractions, multiply the reciprocal).

$$\frac{1/3}{4} = \frac{1}{3} \div 4 = \frac{1}{3} \div \frac{4}{1} = \frac{1}{3} \times \frac{1}{4} = \frac{1}{12}$$

$$\frac{3}{4/5} = \frac{3}{1} \div \frac{4}{5} = \frac{3}{1} \times \frac{5}{4} = \frac{15}{4}$$

2.8 Adding and Subtracting Mixed Numbers

In this section, addition and subtraction of mixed numbers are described.

- To add or subtract mixed numbers:

1. First convert each fraction into improper fractions.
2. Find a common denominator.
3. Add or subtract the numerators.
4. Then place the result over the common denominator.

- **Example:** Add 6 1/2 + 25 3/4.

Convert to improper fractions (see previous section for method).

Convert the first fraction.

$6 \times 2 = 12$

$12 + 1 = 13$

$6\ 1/2 = 13/2$

Convert the second fraction.

$25 \times 4 = 100$

$100 + 3 = 103$

$25\ 3/4 = 103/4$

Find a common denominator for the two improper fractions, 13/2 and 103/4.

4 is a multiple of both 2 and 4.

Multiply each fraction by a fraction with its numerator equal to its denominator such that the result is a common denominator of 4 in each fraction.

Add the numerators and place the result over the common denominator.

$$\frac{13 \times 2}{2 \times 2} + \frac{103 \times 1}{4 \times 1}$$

$$= \frac{26}{4} + \frac{103}{4} = \frac{26 + 103}{4} = \frac{129}{4}$$

Therefore, 6 1/2 + 25 3/4 = 129/4.

2.9 Comparing Fractions: Which is Larger or Smaller?

In this section, methods for comparing the value of fractions are described.

- Two fractions can be directly compared if they have the same denominator. For example, which of the following fractions is larger?

  3/4 or 1/2?

Because 4 is a common multiple of both fractions, it is also a common denominator. To obtain a common denominator, multiply 3/4 by 1/1 and 1/2 by 2/2.

$1/2 \times 2/2 = 2/4$

$3/4 \times 1/1 = 3/4$

The two fractions become:

3/4 and 2/4

Because $3 > 2$,

$3/4 > 2/4$

• To compare several fractions, rather than converting all of the fractions to the same denominator, it may be easier to compare two at a time. Which of the following are larger?

1/8 or 3/4 or 5/6?

First, compare:

1/8 and 3/4

The common denominator is 8.
To obtain a common denominator, multiply 3/4 by 2/2.

$3/4 \times 2/2 = 6/8$

The two fractions become:

6/8 and 1/8

Because $6 > 1$,

$6/8 > 1/8$

Therefore, $3/4 > 1/8$.

Next, compare:

3/4 and 5/6

The common denominator is 12, because 12 is a multiple of both 4 and 6.

Multiply each fraction by a multiplying-fraction so that the new denominators will be 12.

$$\frac{3}{4} \times \frac{3}{3} = \frac{9}{12}$$

$$\frac{5}{6} \times \frac{2}{2} = \frac{10}{12}$$

Because $10 > 9$,

$10/12 > 9/12$

Therefore, $5/6 > 3/4$

Comparing all three fractions, we determined that:

5/6 > 3/4 and 3/4 > 1/8

Therefore, 5/6 > 3/4 > 1/8

- To quickly assess whether one fraction is larger or smaller than another fraction, it may be useful to determine whether each fraction is larger or smaller than 1/2.

- The smaller a numerator is in comparison to its denominator, the smaller will be the value of the fraction.

5/10 > 5/100 > 5/1000 > 5/10,000 > 5/100,000 > 5/1,000,000

58/59 is almost 1.

99/985,236 is approximately one-ten thousandth.

Notes

Chapter 3

Decimals

3.1 Definitions

- The decimal system and decimals are based on tenths or the number 10.

- The digits to the right of the decimal point are called *decimal fractions*.

- A decimal can be expressed in the form of a fraction, and a fraction can be expressed in the form of a decimal.

- Decimals that do not have a digit to the left of the decimal point are written 0.95 or .95. Inserting the zero before the decimal point prevents the viewer from mistaking .95 for 95.

- The digits to the right of the decimal point correspond to tenths, hundredths, thousandths, ten thousandths, hundred thousandths, etc. For example, the digits in the number 48.35679 correspond to:

  4 tens, 8 ones . 3 tenths, 5 hundredths, 6 thousandths,
  7 ten thousandths, 9 hundred thousandths

- The decimal point is always present after the ones digit even though it may not be written.

- **Example:**

  $7 = 7.0 = 7.00 = 7.0000 = 7.0000000$

  $29 = 29.00 = 29.0 = 29.000000000$

  $-36 = -36.0 = -36.00 = -36.00000$

• The following are examples of the equivalent forms of decimals and fractions:

0.1	= 1/10
0.01	= 1/100
0.001	= 1/1,000
0.0001	= 1/10,000
0.00001	= 1/100,000

• Decimals have equivalent forms as fractions. For example, the decimal 0.5 is equal to the fraction 5/10. This can be proven by dividing the fraction 5/10 or $5 \div 10$.

Using the long division format:

$$10\overline{)\,5}$$

Insert the decimal point and a zero in the tenth position.

$$10\overline{)\,5.0}$$

10 divides into 5.0, 0.5 times, then multiply $10 \times 0.5 = 5.0$

$$\begin{array}{r} .5 \\ 10\overline{)\,5.0} \\ \underline{5.0} \\ 00 \end{array}$$

Therefore, 0.5 is equal to 5/10.

• In general, a fraction can be transformed into its decimal equivalent by dividing the numerator by the denominator. (See Section 3.4, "Dividing Decimals.")

$$1/2 = 2\overline{)1} = 2\overline{)1.0} = 2\overline{)\begin{array}{r} .5 \\ 1.0 \end{array}}$$
$$\begin{array}{r} 1.0 \\ \hline 0\,0 \end{array}$$

Therefore, 1/2 = 0.5.

• Proper names that correspond to decimal fractions are used to describe measurements or quantities in various units such as grams, moles, seconds, liters, meters, etc. For example, if a scientist is measuring extremely small amounts of a chemical in grams, proper names for the following decimal quantities are:

$0.1 = 10^{-1} = 100$ milligrams

$0.01 = 10^{-2} = 10$ milligrams

$0.001 = 10^{-3} = 1$ milligram

$0.0001 = 10^{-4} = 100$ micrograms

$0.00001 = 10^{-5} = 10$ micrograms

$0.000001 = 10^{-6} = 1$ microgram

$0.000000001 = 10^{-9} = 1$ nanogram

$0.000000000001 = 10^{-12} = 1$ picogram

$0.000000000000001 = 10^{-15} = 1$ femptogram

3.2 Adding and Subtracting Decimals

• To add or subtract decimals, align the decimal points, then proceed with the addition or subtraction (as if the decimal points are not there).

- To add decimals, arrange in column format, add each column beginning with the right column, and carry over digits to the next larger column as necessary.

- **Example:** Add 389.32 + 65.2 + 2 + 0.05

Arrange in column format, fill in zeros, and add the hundredths' column.

$$\begin{array}{rr} & 389.32 \\ & 65.20 \\ & 2.00 \\ + & .05 \\ \hline & 7 \end{array}$$

Add the tenths' column.

$$\begin{array}{rr} & 389.32 \\ & 65.20 \\ & 2.00 \\ + & .05 \\ \hline & .57 \end{array}$$

Add the ones' column, and carry over the ten.

$$\begin{array}{rl} & \mathit{1} \\ & 389.32 \\ & 65.20 \\ & 2.00 \\ + & .05 \\ \hline & 6.57 \end{array}$$

Add the tens' column, and carry over the hundred.

$$\begin{array}{r} \mathit{11} \\ 389.32 \\ 65.20 \\ 2.00 \\ +\quad .05 \\ \hline 56.57 \end{array}$$

Add the hundreds' column.

$$\begin{array}{r} \mathit{11} \\ 389.32 \\ 65.20 \\ 2.00 \\ +\quad .05 \\ \hline 456.57 \end{array}$$

Therefore, $389.32 + 65.2 + 2 + 0.05 = 456.57$.

- To subtract decimals, arrange in column format, subtract each column beginning with the right column and borrow when necessary from the next larger column. If there are more than two numbers, subtract the first two, then subtract the third from the difference, and so on.

- **Example:** Subtract $23 - 3.89$.

Arrange in column format and subtract the hundredths' column.

$$\begin{array}{r} 23.00 \\ -\quad 3.89 \\ \hline ? \end{array}$$

To borrow from the tenths' column, first borrow from the ones' column. Subtract the hundredths and tenths.

$$\begin{array}{rrcrr} 2 & 2 & .\{9\} & \{10\} \\ - & 3 & .\ 8 & 9 \\ \hline & & .\ 1 & 1 \end{array}$$

To subtract the ones' column, first borrow from the tens'. Subtract the ones and tens.

$$\begin{array}{rrcrr} \{1\} & \{12\} & .\{9\} & \{10\} \\ - & 3 & .\ 8 & 9 \\ \hline 1 & 9 & .\ 1 & 1 \end{array}$$

Therefore, $23 - 3.89 = 19.11$.

3.3 Multiplying Decimals

- To multiply decimals, ignore the decimal points and multiply the numbers, then place the decimal point in the product so that there are the same number of digits to the right of the decimal point in the product as there are in all of the numbers that were multiplied combined.

- For example, if 2.2 and 0.3 are multiplied, because there are two digits to the right of the decimal points in 2.2 and 0.3 (one for each number), there must be two digits to the right of the decimal point in the product. Therefore, $2.2 \times 0.3 = 0.66$.

- **Example:** Multiply 35.268 and 2.5.

There are a total number of four digits to the right of the decimal points in these two numbers, therefore the product will have four digits to the right of the decimal point.

Multiply the 5 in the multiplier with each number in the multiplicand beginning at the right.

$$\begin{array}{r} 35.268 \\ \times \quad 2.5 \\ \hline 176340 \end{array}$$

Multiply the 2 in the multiplier with each number in the multiplicand beginning at the right.

$$\begin{array}{r} 35.268 \\ \times \quad 2.5 \\ \hline 176340 \\ 70536 \\ \hline \end{array}$$

Add the partial products.

$$\begin{array}{r} 35.268 \\ \times \quad 2.5 \\ \hline 176340 \\ 70536 \\ \hline 881700 \end{array}$$

Place the decimal point four digits from the right.

Therefore, $35.268 \times 2.5 = 88.1700$.

3.4 Dividing Decimals

- To divide decimals, arrange the numbers into the long division format, move the decimal point in the divisor to the right until there are no digits to the right of the decimal point, move the decimal point in the dividend the same number of places to the right so that the overall value of the division is unchanged (zeros may be inserted in the dividend as required), then divide using long division. The decimal point will be placed in the quotient above where it moved to in the dividend.

- **Example:** $10 \div 0.5 = ?$

Arrange the numbers into the long division format.

$$0.5\overline{)\,10}^{\;?}$$

Move the decimal point in the divisor and the dividend to the right until there are no digits to the right of the decimal point in the divisor (insert zeros as required).

$$5.\overline{)\,100.}^{\;?}$$

Divide 5 into 10.

$$\begin{array}{r} 2?. \\ 5.\overline{)\,100.} \\ \underline{10} \end{array}$$

Subtract 10 – 10 = 0, and bring down the last 0.

$$\begin{array}{r} 2?.\\ 5.\overline{)\,100.} \\ \underline{10} \\ 000 \end{array}$$

There is nothing for 5 to divide into, so place a 0 above the last 0 in the dividend.

$$\begin{array}{r} 20.\\ 5.\overline{)\,100.} \\ \underline{10} \\ 000 \end{array}$$

Therefore, $10 \div 0.5 = 20$.

3.5 Rounding Decimals

- To round decimals, the last retained digit should either be increased by one or left unchanged according to the following rules:

If the left-most digit to be dropped is less than 5, leave the last retained digit unchanged.

If the left-most digit to be dropped is greater than 5, increase the last retained digit by one.

If the left-most digit to be dropped is equal to 5, leave the last retained digit unchanged if it is even or increase the last retained digit by one if it is odd.

• **Example:** Round the following to the nearest integer.

3.4 rounds to 3.

2.5 rounds to 2.

45.64 rounds to 46.

• Decimals may also be rounded to the nearest tenth, hundredth, thousandth, etc., depending on how many decimal places there are and the accuracy required.

• **Example:** Round the following as specified.

45.689 rounded to the nearest tenth is 45.7.

1.9654 rounded to the nearest hundredth is 1.96.

1.545454 rounded to the nearest thousandth is 1.545.

• When solving complex mathematical or engineering problems, it is important to retain the same number of "significant digits" in the intermediate and final results. The number of decimal places in the resulting numbers should not exceed the number of decimal places in the initial numbers because the resulting numbers cannot be known with greater accuracy than the original numbers. Therefore, depending on the least number of significant digits in the initial numbers, rounding intermediate and final results will be required to maintain the decimal places to ones, tenths, hundredths, thousandths, etc.

• For example, if 45.689 and 1.9654 are added, the accuracy of the result cannot be greater than three decimal places.

$$45.689 + 1.9654 = 47.6544 = 47.654$$

3.6 Comparing the Size of Decimals

• To compare the value of two or more decimals to determine which decimal is a larger or smaller, the following procedure can be applied:

1. Place the decimals in a column.
2. Align the decimal points.
3. Fill in zeros to the right so that both decimals have the same number of digits to the right of the decimal point.
4. The larger decimal will have the largest digit in the greatest column (the farthest to the left).

• **Example:** Compare 0.00025 and 0.000098.

Place the decimals in a column, align the decimal points, and fill in zeros.

0.000250

0.000098

The 2 in the ten-thousandths place is greater than the 9 in the hundred-thousandths place.

Therefore: 0.000250 is larger than 0.000098.

3.7 Decimals and Money

• There is a relationship between money and decimals because the money system is based on decimals. Dollars are represented by whole numbers and cents are represented by tenths and hundredths of a dollar.

- **Example:** \$25.99 = 25 dollars and 99 cents.

- The following are equivalent: Ninety-nine hundredths of a dollar, 99 cents, 99/100 of a dollar, and 0.99 dollars.

- **Examples:**

  10 cents is one tenth or 1/10 of a dollar or 0.1 dollars.

  70 cents is seven tenths or 7/10 of a dollar or 0.7 dollars.

  1 cent is one hundredth or 1/100 of a dollar or 0.01 dollars.

  6 cents is six hundredths or 6/100 of a dollar or 0.06 dollars.

- The amount, \$0.50 is 1/2 of a dollar. Similarly, the decimal 0.5 equals the fraction 1/2.

- The word cent refers to one hundred, (remember a *cent*ury has 100 years), and in money, cents refers to the number of hundredths.

Notes

Chapter 4

Percentages

4.1 Definitions

- Percent is defined as a rate or proportion per hundred, one one-hundredth, or 1/100.
- The symbol for a percent is %.
- A percent is a form of a fraction with its denominator equal to 100. To remember that per*cent* is per one-hundred, think of a *cent*ury, which has 100 years.
- A percent can be converted to a fraction or decimal. A percent can be reduced and manipulated like a fraction or decimal.
- For example, seventy-five percent = 75% = 75/100 = 0.75.

- **Examples:** Common percents are:

1/4	=	0.25	=	25%
1/2	=	0.50	=	50%
3/4	=	0.75	=	75%
1/3	=	0.33...	=	33 1/3%
2/3	=	0.66...	=	66 2/3%
1/5	=	0.20	=	20%
1/100	=	0.01	=	1%
1/50	=	0.02	=	2%
1/25	=	0.04	=	4%
1/10	=	0.10	=	10%
2/5	=	0.40	=	40%
3/5	=	0.60	=	60%
4/5	=	0.80	=	80%
1	=	1.00	=	100%
6/5	=	1.20	=	120%
5/4	=	1.25	=	125%
3/2	=	1.50	=	150%

4.2 Figuring Out the Percents of Numbers

- Percents of a number can be determined using the definition of a percent and multiplying the number by the amount of the percent and 1/100. The following examples represent percents of the number nine:

$$1\% \text{ of } 9 = 1 \times \frac{1}{100} \text{ of } 9$$

$$= (1)(\frac{1}{100})(9) = (9)(\frac{1}{100}) = \frac{9}{100} = 0.09$$

$$10\% \text{ of } 9 = 10 \times \frac{1}{100} \text{ of } 9$$

$$= (10)(\frac{1}{100})(9) = (90)(\frac{1}{100}) = \frac{90}{100} = 0.9$$

$$100\% \text{ of } 9 = 100 \times \frac{1}{100} \text{ of } 9$$

$$= (100)(\frac{1}{100})(9) = (900)(\frac{1}{100}) = \frac{900}{100} = 9$$

$$1{,}000\% \text{ of } 9 = 1{,}000 \times \frac{1}{100} \text{ of } 9$$

$$= (1{,}000)(\frac{1}{100})(9) = (9{,}000)(\frac{1}{100}) = \frac{9{,}000}{100} = 90$$

- Percents of a number can also be determined by moving the decimal point according to the percent amount. The following examples represent percents of the number nine:

  1% of 9. Move the decimal to the left two digits. = 0.09

  10% of 9. Move the decimal to the left one digit. = 0.9

  100% of 9. The decimal does not move. = 9

  1000% of 9. Move the decimal to the right one digit. = 90

  10,000% of 9. Move the decimal to the right two digits = 900

  100,000% of 9. Move the decimal to the right three digits. = 9,000

- The following are examples of percents of numbers:

What is 6 percent of 30?

$6 \times 1/100$ of $30 = 6 \times 1/100 \times 30$

$= 6/100 \times 30 = 180/100 = 1.80$

What is 5 percent of 20?

$5 \times 1/100$ of $20 = 5 \times 1/100 \times 20 = 100/100 = 1.00$

What is 20 percent of 80?

$20 \times 1/100$ of $80 = 20 \times 1/100 \times 80 = 1600/100 = 16$

• The examples above can also be solved using an alternative method.

What is 6 percent of 30?

Because 1% of 30 is 0.30, and 6% is 6 times 1%, then $0.30 \times 6 = 1.80$.

What is 5 percent of 20?

Because 10% of 20 is 2, and 5% is half of 10%, then half of 2 is 1.

What is 20 percent of 80?

Because 10% of 80 is 8, and 20% is 2 times 10%, then $8 \times 2 = 16$.

4.3 Adding, Subtracting, Multiplying, and Dividing Percents

• To add or subtract percents, simply add or subtract as with integers but keep track of the percent.

• **Example:** Add or subtract the following:

$26\% + 4\% = 30\%$

$158\% - 6\% = 152\%$

$3\% + 16\% + 4\% = 23\%$

• For example, if you have 50% of your assets in stocks and 20% of your assets in bonds, then 70% of your assets are in stocks and bonds.

50% in stocks + 20% in bonds = 70% in stocks and bonds

- To multiply and divide percents:

1. Convert percents to decimal form by dividing by 100 (move decimal two places to left).
2. Multiply or divide the decimals (as described in Chapter 3).
3. Convert back to percent by multiplying by 100 (move decimal point two places to right).

- Multiply or divide the following examples:

$25\% \times 5\% = 0.25 \times 0.05 = 0.0125$

Convert to %: $0.0125 \times 100 = 1.25\%$

$75\% \div 100\% = 0.75 \div 1.00 = 0.75$

Convert to %: $0.75 \times 100 = 75\%$

$80\% \div 25\% = 0.8/0.25 = 3.2$

Convert to %: $3.2 \times 100 = 320\%$

6% of $30 = 6\% \times 30 = 0.06 \times 30 = 1.80$

5% of $20 = 5\% \times 20 = 0.05 \times 20 = 1$

20% of $80 = 20\% \times 80 = 0.2 \times 80 = 16$

20% of 70% of something $= 0.2 \times 0.7 = 0.14 = 14\%$ of something

5 is what percent of 20? $5/20 = 0.25 = 25\%$

4.4 Percent Increase and Decrease (Percent Change)

In this section, percent change and discount are discussed.

• To determine the percent that a number has increased or decreased, the following equation is used:

$$\frac{\text{amount of increase or decrease (change)}}{\text{original amount}} = \frac{?}{100}$$

This equation can also be written:

$$\frac{\text{amount of increase or decrease (change)}}{\text{original amount}}(100) = \text{percent change}$$

• **Example:** If the price of a house is discounted from \$250,000 to \$200,000, what is the percent decrease or discount in the price?

$$\frac{\$250{,}000 - \$200{,}000}{\$250{,}000}(100) = \frac{\$50{,}000}{\$250{,}000}(100)$$

$$= (0.2)(100) = 20\%$$

• **Example:** If the owners of a house are willing to discount the price of \$150,000 by 6%, how much will the house cost? (See Chapter 3, "Solving Simple Algebraic Equations," in *Algebra*, the second book of the *Master Math* series, for explanations of techniques used below.)

$$\frac{\$150{,}000 - \text{discount price}}{\$150{,}000} = \frac{6\%}{100}$$

$$\frac{\$150{,}000 - \text{discount price}}{\$150{,}000} = 0.06$$

Multiply both sides by \$150,000.

$$\$150{,}000 - \text{discount price} = (0.06)(\$150{,}000)$$

$$\$150{,}000 - \text{discount price} = \$9{,}000$$

Subtract \$9,000 from both sides, and add discount price to both sides.

$$\$150{,}000 - \$9{,}000 = \text{discount price}$$

$$\$141{,}000 = \text{discount price}$$

To verify the result, use the discount price and calculate the percent discount given (6%).

$$\frac{\$150{,}000 - \$141{,}000}{\$150{,}000}(100) = ? = 6\%$$

4.5 Simple and Compound Interest

• *Simple interest* is generally computed annually, therefore the time it will take to earn a certain percent on money invested in a simple interest account is a year.

• **Example:** How much simple interest will \$5,000 earn in a year and in 6 months at a rate of 4%?

$\$5{,}000 \times 4\% = (\$5{,}000)(0.04) = \$200$ will be earned in a year.

In the first 6 months, \$5,000 will earn:

$$\$200 \times (6 \text{ months})/(12 \text{ months}) = \$200 \times 1/2 = \$100$$

Therefore, in 6 months the \$5,000 becomes \$5,100 and in one year the \$5,000 becomes \$5,200.

• *Compound interest* is compounded periodically during the year. To determine the compound interest, divide the interest rate by the number of times it is compounded in the year and apply it during each period.

• For example, if the interest rate is 4% and it is compounded semiannually, then 2% is applied to the principle every 6 months. If 4% is compounded quarterly then 1% is applied to the principle every 3 months or four times a year.

• **Example:** How much interest will \$5,000 earn in six months and in a year at 4% interest compounded semi-annually?

Divide the interest by the number of times it is compounded.
$4\% \div 2 = 2\%$

Apply 2% interest for the first 6-month period.
$\$5{,}000 \times 2\% = (\$5{,}000)(0.02) = \$100$ earned in 6 months.

The second 6-month period begins with \$5,100.

Apply 2% interest again for the second 6-month period.
$\$5{,}100 \times 2\% = (\$5{,}100)(0.02) = \$102$ earned in the second 6-month period.

In 6 months the \$5,000 becomes \$5,100

And in one year the \$5,000 becomes \$5,202

• Note that compound interest earns slightly more than simple interest for the same rate.

Notes

Chapter 5

Converting Percentages, Fractions, and Decimals

5.1 Converting Fractions to Percents

- To convert a fraction to a percent, divide the numerator by the denominator, then multiply by 100.

- **Example:** Convert 3/5 into a percent.

Divide the numerator by the denominator using long division.

$$5\overline{)\,3.0}^{\;?} \quad = \quad \begin{array}{r} .6 \\ 5\overline{)\,3.0} \\ \underline{3.0} \\ 0.0 \end{array}$$

Multiply by 100. $0.6 \times 100 = 60\%$

Therefore, $3/5 = 60\%$.

- A fraction can also be converted to a percent by determining an equivalent form of the fraction that has a denominator of 100.

- **Example:** Convert 3/5 into a percent.

$$\frac{3}{5} = \frac{6}{10} = \frac{60}{100} = 60\%$$

5.2 Converting Percents to Fractions

- To convert a percent to a fraction, divide by 100.

- **Example:** Convert 60% into a fraction.

Divide by 100.

$$60\% = \frac{60}{100} = \frac{6}{10} = \frac{3}{5}$$

Therefore, 60% = 3/5

5.3 Converting Fractions to Decimals

- To convert a fraction to a decimal, divide the numerator by the denominator.

- **Example:** Convert 3/5 into a decimal.

Divide the numerator by the denominator using long division.

$$5\overline{)\,3.0}^{\;?} \quad = \quad \begin{array}{r} .6 \\ 5\overline{)\,3.0} \\ \underline{3.0} \\ 0.0 \end{array}$$

Therefore, 3/5 = 0.6.

5.4 Converting Decimals to Fractions

- To convert a decimal to a fraction, place the decimal-fraction over the tenth, hundredth, thousandth, ten-thousandth, hundred-thousandth, etc., that it corresponds to, then reduce the fraction.

• **Example:** Convert the following decimals into their fractional form.

$$0.02 = \frac{2}{100} = \frac{1}{50}$$

$$0.5 = \frac{5}{10} = \frac{1}{2}$$

$$0.6 = \frac{6}{10} = \frac{3}{5}$$

$$0.0056 = \frac{56}{10{,}000} = \frac{7}{1{,}250}$$

$$0.446 = \frac{446}{1{,}000} = \frac{223}{500}$$

To check the results, divide each fraction.

• Remember, 0.6 is 6 tenths, 0.06 is 6 hundredths, and so on.

• Consider the following decimals in their fractional form.

0.1 = 1/10

0.01 = 1/100

0.001 = 1/1,000

0.0001 = 1/10,000

0.00001 = 1/100,000

0.000001 = 1/1,000,000

5.5 Converting Percents to Decimals

- To convert a percent to a decimal, divide by 100.

- Note that dividing by 100 is equivalent to moving the decimal point two places to the left.

- **Example:** Convert 25% to its decimal form.

Divide 25 by 100.

$$25\% = \frac{25}{100} = 0.25$$

- The above example is verified using long division.

$$25\% = \frac{25}{100} = 25 \div 100 = ?$$

$$100\overline{)\,25}\quad ?$$

Insert the decimal point and a zero after 25, divide 100 into 25.0, multiply the result by 100, and subtract.

$$\begin{array}{r} .2 \\ 100\overline{)\,25.0} \\ \underline{20.0} \end{array}$$

$$\begin{array}{r} .2 \\ 100\overline{)\,25.0} \\ \underline{20.0} \\ 05.0 \end{array}$$

Insert another zero, divide 100 into 5.00, multiply the result by 100, and subtract.

$$\begin{array}{r@{\,}r} & .25 \\ 100 \big) & \overline{25.00} \\ & \underline{20.0} \\ & 05.00 \\ & \underline{05.00} \end{array}$$

$$\begin{array}{r@{\,}r} & .25 \\ 100 \big) & \overline{25.00} \\ & \underline{20.0} \\ & 05.00 \\ & \underline{05.00} \\ & 00.00 \end{array}$$

Therefore, 25% is equivalent to 0.25.

5.6 Converting Decimals to Percents

- To convert a decimal to a percent, multiply by 100.

- Note that multiplying by 100 is equivalent to moving the decimal point two places to the right.

- **Example:** Convert the decimal 0.25 to its percent form.

  Multiply 0.25 by 100.

  $0.25 \times 100 = 25\%$

- Two final notes:

If two decimals are multiplied, the product will have the same number of digits to the right of its decimal point as in the multiplicand and multiplier combined.

When converting between fractions, percents, and decimals, think about whether the result seems reasonable. For example, if the fraction 3/4 is converted to a percent and the result is 0.0075%, it does not seem reasonable that 0.0075% could be equivalent to three-fourths. Instead, 75% does seem reasonable.

Notes

Chapter 6

Ratios, Proportions, and Variation

6.1 Definitions

In this section, ratios and proportions are defined.

• *Ratios* depict the relation between two similar values with respect to the number of times the first contains the second. A ratio represents a comparison between two quantities. For example, if the ratio between apples and oranges in a fruit bowl is 3 to 2, then for every 3 applies there are 2 oranges.

• Ratios are written 3 to 2, 3:2, and 3/2.

• The difference between ratios and fractions is that fractions represent part-per-whole, and ratios represent part-per-part.

• *Proportions* are a comparative relation between the size, quantity, etc., of objects or values. Mathematical proportions represent two equal ratios. For example, the following are equivalent:

$$1/2 = 2/4,\ 1{:}2 = 2{:}4,\ 1/2 \text{ is proportional to } 2/4,\ 1{:}2{::}2{:}4$$

• Note that in a proportion, the product of the "extreme" outer numbers equals the product of the "mean" middle numbers. 1:2::2:4 can be written $1 \times 4 = 2 \times 2$.

• When equivalent fractions, such as $1/2 = 2/4$, are cross-multiplied (by multiplying the opposite numerators with the opposite denominators), the two products are always equal. This principle is useful when determining the quantity of one of the items when the ratio is known.

• For example, how many apples are there in a mixture containing 10 oranges, if the ratio is 3 apples to 2 oranges? To solve this, set up the proportion:

$$\frac{3 \text{ apples}}{2 \text{ oranges}} = \frac{? \text{ apples}}{10 \text{ oranges}}$$

Using cross-multiplication the equation becomes:

$3 \times 10 = 2 \times ?$

$30 = 2 \times ?$

Divide both sides by 2.

$30/2 = 1 \times ?$

$15 = ?$

Therefore, there are 15 applies in this mixture.

6.2 Comparing Ratios to Fractions and Percents

- The difference between ratios and fractions is that fractions represent part-per-whole, and ratios represent part-per-part.
- Ratios can be expressed in the form of a fraction, (or in the form of division).
- The definition of a ratio differs from that of a fraction but all the rules for manipulating fractions apply to ratios.
- The ratio 1 to 2 (or 1:2 or 1/2) can be expressed as 50% or 0.50.

- **Example:** If Tom ate 3 pies for every 5 pies Ted ate, what is the ratio of pies eaten by Tom and Ted?

<u>Tom</u> : <u>Ted</u>
(3 pies) / (5 pies)

Tom ate 3/5 the number of pies that Ted ate.

Tom ate $3/5 = 0.60 = 60\%$ of the number of pies that Ted ate.

- If a ratio of two values is known, the percent that one of the values is of the total can be determined by adding the parts to get the whole, placing the one value over the whole, and multiplying by 100.

- **Example:** What percent of the total number of pies did Tom and Ted eat?

The ratio is (3 pies) to (5 pies), which represents part/part.

And the percent of the total number is (part/whole)(100).

If the total number or whole is 3 pies + 5 pies = 8 pies,

then, the percent of the total number of pies Tom ate is 3 pies/8 pies $= 3/8 = 0.375$ and $0.375 \times 100 = 37.5\%$

and, the percent of the total number of pies Ted ate is 5 pies/8 pies $= 5/8 = 0.625$ and $0.625 \times 100 = 62.5\%$

6.3 Variation and Proportion

- Variation describes how one quantity varies or changes as another quantity varies or changes.

- If as one number increases or decreases, another number increases or decreases simultaneously in the same proportion, the numbers are in *direct variation* or *direct proportion* to each other. The equation for direct variation or direct proportion is:

$$y = kx$$

The k represents some constant number, and x and y represent numbers that vary directly and proportionately. As x increases, y increases. As x decreases, y decreases.

- If as one number increases or decreases, another number decreases or increases simultaneously in the same proportion, the numbers are in *indirect variation* and are said to be *inversely proportional*. The equation for indirect variation representing inversely proportional quantities is:

$$xy = k \quad \text{which can be rearranged as } y = k/x$$

The k represents some constant number, and x and y represent numbers that vary indirectly and proportionately. As x increases, y decreases. As x decreases, y increases.

Notes

Chapter 7

Powers and Exponents

7.1 Definition of the Exponent or Power

• An exponential expression represents a number that is multiplied by itself the number of times as the exponent or power defines.

• The following exponential expression represents 2 raised to the 6th power:

$$2^6 = 2 \times 2 \times 2 \times 2 \times 2 \times 2 = 64$$

2 is called the *base* and 6 is the *exponent* or *power*.

• Examples of exponential expressions are 2^6, 25^7, and 219^{65}.

• If a positive whole number is raised to a power greater than one, the result is a larger number.

• If a number (except zero) is raised to the 1st power, the result is the number itself. $3^1 = 3$.

• If a number is raised to the zero power, the result is 1.

$$3^0 = 1$$

$$a^0 = 1$$

$$a = \text{any number except zero}$$

$$0^0 = \text{undefined}$$

7.2 Negative Exponents and Bases

In this section, negative numbers raised to even and odd powers, and positive and negative numbers raised to negative powers are presented.

- If a negative number is raised to a positive even power, the result is a positive number.

- **Example:**

$$(-2)^6 = (-2)(-2)(-2)(-2)(-2)(-2) = 64$$

- If a negative number is raised to a positive odd power, the result is a negative number.

- **Example:**

$$(-2)^5 = (-2)(-2)(-2)(-2)(-2) = -32$$

- If a positive or negative number is raised to a negative power, it can be equivalently written as one over that number raised to the positive power.

- **Example:**

$$2^{-6} = \frac{1}{2^6} = \frac{1}{2} \times \frac{1}{2} \times \frac{1}{2} \times \frac{1}{2} \times \frac{1}{2} \times \frac{1}{2} = \frac{1}{64}$$

$$(-2)^{-6} = \frac{1}{(-2)^6} = \frac{1}{-2} \times \frac{1}{-2} \times \frac{1}{-2} \times \frac{1}{-2} \times \frac{1}{-2} \times \frac{1}{-2} = \frac{1}{64}$$

7.3 Multiplying Exponential Expressions with the Same Base

- To multiply exponential expressions with the same base, simply add the powers.

- **Example:**

$$3^2 \times 3^4 = 3^{2+4} = 3^6 = 3 \times 3 \times 3 \times 3 \times 3 \times 3 = 729$$

This example can be equivalently written:

$$(3 \times 3) \times (3 \times 3 \times 3 \times 3) = 9 \times 81 = 729$$

- The following is one of the laws of exponents (a, b, and c represent numbers):

$$a^b \times a^c = a^{b+c}$$

7.4 Multiplying Exponential Expressions with Different Bases

- To multiply exponential expressions with different bases, each exponent must be expressed individually, then multiplied. The powers cannot be added directly.

- **Example:**

$$3^2 \times 2^4 = (3 \times 3) \times (2 \times 2 \times 2 \times 2) = 9 \times 16 = 144$$

$a^b \times d^c = a^b d^c$ Where a, b, c, and d represent numbers.

- To multiply exponential expressions with different bases but the same power, the following law of exponents applies (a, b, and c represent numbers):

$$a^c \times b^c = (a \times b)^c$$

Which can also be written:

$$a^c b^c = (ab)^c$$

7.5 Dividing Exponential Expressions with the Same Base

- To divide exponential expressions with the same base, subtract the powers.

- **Example:**

$3^4 \div 3^2 = (3^4)/(3^2) = 3^{4-2} = 3^2 = 3 \times 3 = 9$

This example can be equivalently written:

$$3^4 \div 3^2 = \frac{3^4}{3^2} = \frac{3 \times 3 \times 3 \times 3}{3 \times 3} = \frac{81}{9} = 9$$

- **Example:**

$3^2 \div 3^4 = (3^2)/(3^4) = 3^{2-4} = 3^{-2} = 1/3^2 = 1/9$

- Consider the following examples (a, b, and c represent numbers):

$a^3b^2/a^2b = (aaabb)/(aab) = a^{3-2}b^{2-1} = a^1b^1 = ab$

$a^2b/a^3b^2 = (aab)/(aaabb)$

$= a^{2-3}b^{1-2} = a^{-1}b^{-1} = (ab)^{-1} = 1/(ab)$

$a^b \div a^c = (a^b)/(a^c) = a^{b-c}$

$a^b \div a^b = (a^b)/(a^b) = a^{b-b} = a^0 = 1$

$(a+b)^b \div (a+b)^b = (a+b)^b/(a+b)^b = (a+b)^{b-b} = (a+b)^0 = 1$

- A base with an exponent of zero is equal to 1.

$54^0 = 1$

$a^0 = 1$ Where *a* represents a number.

0^0 is undefined.

7.6 Dividing Exponential Expressions with Different Bases

• To divide exponential expressions with different bases, each exponent must be expressed individually, then divided. The powers cannot be subtracted directly.

• **Example:**

$$\frac{3^3}{2^2} = \frac{3 \times 3 \times 3}{2 \times 2} = \frac{27}{4} = 6.75$$

7.7 Raising Exponential Expressions to a Power

• To express an exponential expression raised to a power, multiply the powers.

• **Example:**

$$(3^2)^3 = 3^{2\times 3} = 3^6 = (3 \times 3 \times 3 \times 3 \times 3 \times 3) = 729$$

Equivalently:

$$(3^2)^3 = (3^2)(3^2)(3^2) = (3 \times 3)(3 \times 3)(3 \times 3) = 9 \times 9 \times 9 = 729$$

Equivalently:

$$(3^2)^3 = (3 \times 3)^3 = (9)^3 = (9) \times (9) \times (9) = 729$$

• **Example:** Evaluate $(-y^2)^2$, (y represents a number).

$$(-y^2)^2 = (-y^2) \times (-y^2) = (-1 \times y \times y) \times (-1 \times y \times y)$$

$$= 1 \times y \times y \times y \times y = y^4$$

Equivalently:

$$(-y^2)^2 = (-1 \times y^2)^2 = (-1)^2 \times (y^2)^2 = 1 \times y^{2*2} = 1 \times y^4 = y^4$$

- The following is one of the laws of exponents, (a, b, and c represent numbers):

$$(a^b)^c = a^{b \times c} = a^{bc}$$

7.8 Distributing Exponents into Parentheses

- If numbers or variables inside parentheses are multiplied or divided and have an exponent outside the parentheses, the exponent can be *distributed* to each number or variable inside the parentheses.

- **Example:**

$$(3 \times 2)^2 = (3^2 \times 2^2) = 9 \times 4 = 36$$

- Consider the following examples (a, b, and c represent numbers):

$$(ab)^2 = (a^2 \times b^2) = a^2b^2$$

$$(ab)^c = a^cb^c$$

$$(2b)^2 = (2^2 \times b^2) = 4b^2$$

$$(2/3)^2 = (2^2/3^2) = 4/9$$

$$(a/b)^2 = (a^2/b^2) = a^2/b^2$$

• If numbers or variables inside parentheses are added or subtracted, the exponent outside cannot be *distributed* as above. Instead:

$$(a + b)^2 = (a + b)(a + b)$$

$$= a^2 + ab + ab + b^2 = a^2 + 2ab + b^2$$

(See multiplication of binomials in Chapter 5 of the second book in the *Master Math* series, *Algebra*.)

7.9 Addition of Exponential Expressions

• To add exponential expressions whether the base is the same or different, express each exponential expression individually, then add. The powers cannot be added directly.

• **Example:**

$$3^2 + 2^4 = (3 \times 3) + (2 \times 2 \times 2 \times 2) = 9 + 16 = 25$$

$$3^2 + 3^4 = (3 \times 3) + (3 \times 3 \times 3 \times 3) = 9 + 81 = 90$$

7.10 Subtraction of Exponential Expressions

• To subtract exponential expressions whether the base is the same or different, express each exponential expression individually, then subtract. The powers cannot be subtracted directly.

• **Example:**

$$2^4 - 3^2 = (2 \times 2 \times 2 \times 2) - (3 \times 3) = 16 - 9 = 7$$

$$3^4 - 3^2 = (3 \times 3 \times 3 \times 3) - (3 \times 3) = 81 - 9 = 72$$

7.11 Exponential Expressions Involving Fractions

- To express fractions containing exponential expressions, apply the principles in the preceding sections in this chapter to each expression within the fraction. For example, if parentheses surround numbers or variables multiplied together, the exponent can be distributed to each number or variable within the parentheses.

- **Example:** Express the following fraction.

$$\frac{(a^b)^c}{(2b)^2} = \frac{a^{b \times c}}{(2^2 \times b^2)} = \frac{a^{bc}}{4b^2}$$

- A number raised to a negative power is equivalent to one over that number.

$$3^{-1} = \frac{1}{3} \text{ and } 3^{-6} = \frac{1}{3^6}$$

- If a positive fraction is raised to a power, and the numerator is smaller than the denominator, the value of the resulting fraction is less.

$(1/2)^2 = (1/2)(1/2) = 1/4$ Where $1/4 < 1/2$.

- If a positive fraction is raised to a power, and the numerator is larger than the denominator, the value of the resulting fraction is greater.

$(5/2)^2 = (5/2)(5/2) = 25/4$ Where $25/4 > 5/2$.

Remember, $5/2 = 10/4$.

- Consider the following examples with fractional exponents:

$$3^{1/2} = \sqrt{3}$$

$$3^{2/3} = (3^{1/3})^2 = (3^2)^{1/3}$$

Notes

Chapter 8

Logarithms

8.1 Definition of the Logarithm

- The logarithm of a number is the power to which a base number must be raised to equal that number.

- A logarithm is the inverse of an exponent.
- Each exponent has an inverse logarithm.
- Consider the following exponential equation (x and y represent numbers):

$$x = a^y$$

Where *a* is an integer and is the base of the exponent.

The inverse of this equation is:

$$y = \log_a x$$

Where *a* is the base of the logarithm.

The inverse is obtained by taking the base *a* logarithm of both sides of the exponential equation.

$$x = a^y$$

$$\log_a(x) = \log_a(a^y)$$

Because $\log_a(a)$ cancels, the equation becomes:

$$\log_a(x) = y$$

- **Example:** Convert the exponential equation $2^4 = 16$ to a logarithmic equation.

Because the exponent has a base of 2, take the base 2 logarithm of both sides.

$$\log_2(2^4) = \log_2(16)$$

Because $\log_2(2)$ cancels, the equation becomes:

$$4 = \log_2(16)$$

8.2 Common (Base Ten) and Natural Logarithm

- The two most common logarithms are the common logarithm, also called the base ten logarithm, and the natural logarithm.

- The common or base ten logarithm is written:

$$\log_{10}x \quad \text{or simply} \quad \log x$$

- The natural logarithm is written:

$$\log_e x \quad \text{or} \quad \ln x$$

Where $e = 2.71828182846$.

- The relationship between the common logarithm and the natural logarithm can be written:

$$\log_e x = (2.3026)\log_{10}x$$

Or equivalently:

$$\ln x = (2.3026)\log x$$

8.3 Solving Equations with Logarithms or Exponents

- To solve logarithmic equations, take the inverse or exponent.

- **Example:** Solve the equation $y = \log x$ for x.

Because the base is 10, raise both sides of the equation by base 10.

$$10^y = 10^{(\log x)}$$

Because $10^{(\log_{10})}$ cancels itself, $10^{(\log x)} = x$

The equation becomes:

$$10^y = x$$

- **Example:** Solve the equation $y = \ln x$ for x.

Because the base is e, raise both sides of the equation by e to isolate x.

$$e^y = e^{(\ln x)}$$

Because $e^{\ln}$ cancels itself, $e^{(\ln x)} = x$

The equation becomes:

$$e^y = x$$

- To solve exponential equations, take the inverse or logarithm.

• **Example:** Solve the equation $y = 10^x$ for x.

Because the base is 10, take $\log_{10}$ of both sides of the equation.

$$\log y = \log(10^x)$$

Because $\log_{10}(10)$ cancels itself, $\log(10^x) = x$

The equation becomes:

$$\log y = x$$

• **Example:** Solve the equation $y = e^x$ for x.

Because the base is e, take ln of both sides of the equation.

$$\ln y = \ln(e^x)$$

Because ln(e) cancels itself, $\ln(e^x) = x$

The equation becomes:

$$\ln y = x$$

8.4 Exponential Form and Logarithmic Form

In this section, examples of computed logarithms and exponents are presented.

- **Exponential Form**

  $10^0 = 1$

  $10^1 = 10$

  $10^2 = 100$

  $10^3 = 1{,}000$

  $10^4 = 10{,}000$

  $10^5 = 100{,}000$

  $e^1 = 2.71828182846$

  $e^\infty = \infty$

  $e^{-\infty} = 0$

  $e^0 = 1$

  $e^{2.30258509299} = 10$

- **Logarithmic Form**

  $\log 1 = 0$

  $\log 10 = 1$

  $\log 100 = 2$

  $\log 1{,}000 = 3$

  $\log 10{,}000 = 4$

  $\log 100{,}000 = 5$

  $\ln (2.71828182846) = \ln e = 1$

  $\ln \infty = \infty$

  $\ln 0 = -\infty$

  $\ln 1 = 0$

  $\ln 10 = 2.30258509299$

• To approximate or compute numerical values of logarithms or exponents, refer to tables contained in mathematical handbooks and selected algebra books, or use the function keys of technical calculators.

8.5 Laws of Logarithms: Addition, Subtraction, Multiplication, Division, Power, and Radical

• The following are laws of logarithms (a represents any base and n is a number):

$$\log_a(x \times y) = \log_a x + \log_a y$$

$$\log_a(x/y) = \log_a x - \log_a y$$

$$\log_a(x^n) = (n)\log_a x$$

$$\log_a(\sqrt[n]{x}) = (1/n)\log_a x \quad \text{Where} \quad \sqrt[n]{x} = x^{1/n}$$

Also, note that: $\log_b a = 1/\log_a b$

Where a and b represent numbers.

Because *a* represents any base, these laws apply to base e (natural logarithms) or base ten (common logarithms). Proofs of the laws of logarithms can be found in selected algebra books.

8.6 Examples: The Richter Scale, pH, and Radiometric Dating

In this section, applications of logarithmic and exponential equations for the Richter scale, the pH scale, and radiometric dating are presented.

- The Richter scale is a logarithmic scale that ranges from 1 to 10 and is used for indicating the intensity of an earthquake. The Richter scale was originated in 1931 by K. Wadati in Japan and was developed by Charles Richter in California in 1935. The Richter scale is a quantitative scale that measures wave amplitudes generated from ground shaking using a seismograph. Richter defined the magnitude of a local earthquake as the logarithm to base ten of the maximum seismic-wave amplitude (in thousandths of a millimeter) recorded on a standard seismograph at a distance of 100 kilometers from the earthquake epicenter. Using the logarithmic scale, for each one unit increase in magnitude, the amplitude of the earthquake waves increase ten times.

- pH expresses the acidity or basicity of a solution in terms of the pH (potency of hydrogen) scale. pH is defined as the negative of the logarithm of the hydrogen (or hydronium) ion concentration in an aqueous (water is the solvent) solution.

$$\text{pH} = -\log[\text{H}^+] \quad \text{or} \quad \text{pH} = -\log[\text{H}_3\text{0}^+]$$

For example, pure water that has a neutral pH of 7 has a hydrogen ion concentration of 1×10^{-7} M or 10^{-7} M (M is moles per liter). The logarithm of 10 raised to the power –7 is equal to –7.

$$\text{pH} = -\log[\text{H}^+] = -\log[10^{-7}] = -(-7) = 7$$

In an acidic solution, the H^+ concentration is greater than 10^{-7} M and less than the OH^- concentration. For example, an acidic solution with 1×10^{-4} M of H^+ has a pH of 4.

$$\text{pH} = -\log[\text{H}^+] = -\log[10^{-4}] = -(-4) = 4$$

In a basic solution, the H^+ concentration is less than 10^{-7} M and less than the OH^- concentration. For example, a basic solution with 3.5×10^{-10} M of H^+ has a pH of 9.46.

$$pH = -\log[H^+] = -\log[3.5 \times 10^{-10}]$$

$$= -(\log[3.5] + \log[10^{-10}])$$

$$= -(0.54 - 10) = -(-9.46) = 9.46$$

- Radioactive substances are subject to exponential decay. The decay of radioactive substances is used to estimate the age of objects. This is called radiometric dating or radioactive dating. For example, carbon-14 is used to estimate the age of plants and animals that were once alive by comparing the ratio C^{14}/C^{12}. (C–12 is the more common form of carbon.) Carbon-14 is produced in the atmosphere when high-energy neutrons (produced from cosmic rays colliding with atmospheric molecules) collide with the nuclei of nitrogen-14 in the air. The carbon-14 then reacts with oxygen to form carbon-14-carbon dioxide.

The carbon-14-carbon dioxide is incorporated into plants through photosynthesis and into animals through ingestion of plants. While a plant or animal is alive, the content of carbon-14 in its system is assumed to remain constant in small but measurable quantities. After the plant or animal has died, carbon-14 can no longer be incorporated by photosynthesis or ingestion, and the carbon-14 that is present begins to decay back to nitrogen-14 by emission of beta radiation (electrons).

The half-life of carbon-14 is 5,730 years; therefore, after 5,730 years half of the original amount of carbon-14 in a given sample will have decayed back to nitrogen-14. (Note that radiometric or radioactive dating techniques assume: (a) the rate of radioactive decay has remained constant over time; (b) the concentration of the radioactive substance that is naturally abundant today is equivalent to the concentration that was naturally abundant in the past; and (c) the substance did not interact with the environment so that there were no original atoms or decay products added or removed.)

The equation for the mass of a substance at any time, t, can be derived as follows. If the half-life of a radioactive substance is $t_{1/2}$ and its initial mass is c, then, in $t_{1/2}$ units of time (years), the mass will reduce to c/2. The mass is given by $y(t_{1/2}) = ce^{-kt_{1/2}}$, therefore:

$$ce^{-kt_{1/2}} = c/2$$

Canceling the c's and taking the natural logarithm of both sides, the equation becomes:

$$\ln(e^{-kt_{1/2}}) = \ln(1/2)$$

Because ln e cancels itself, and because $\ln(1/2) = -\ln 2$, the equation becomes:

$$-kt_{1/2} = -\ln 2$$

Canceling the negative signs on both sides, and dividing both sides by k, the equation becomes:

$$t_{1/2} = (\ln 2)/k = (0.693)/k$$

So:

$$t_{1/2} = (0.693)/k$$

Multiplying both sides by k, and then dividing both sides by $t_{1/2}$, the equtiion becomes:

$$k = (0.693)/t_{1/2}$$

Therefore, the mass of a substance at any time t is:

$$y(t) = ce^{-kt} = ce^{-(0.693)(t/t_{1/2})}$$

(See Chapter 3 of *Algebra,* the second book in the *Master Math* series, for an explanation of solving equations using algebraic techniques.)

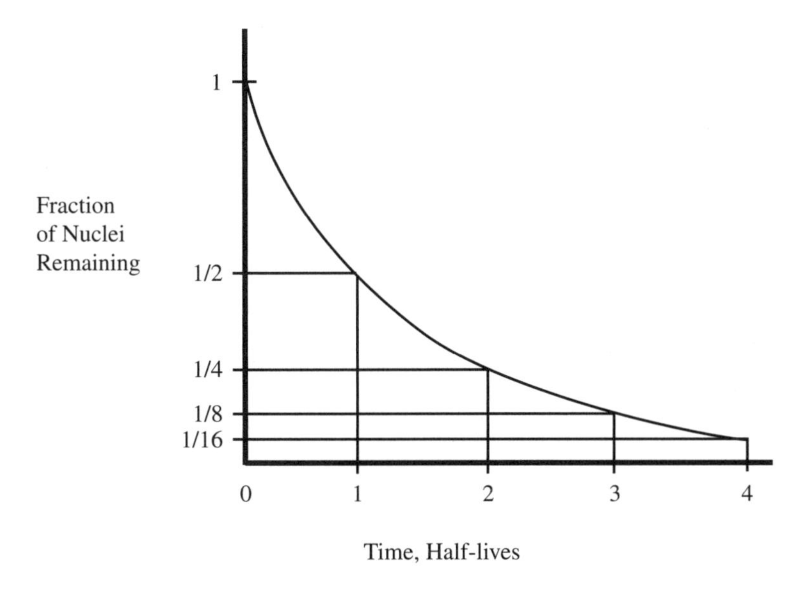

The number of nuclei remaining is halved during each half-life. The greatest decrease occurs during the first half-life and the amount of decrease lessens with each succeeding half-life.

Notes

Chapter 9

Roots and Radicals

9.1 Definitions

• A *radical* is written using the radical symbol, $\sqrt{\ }$. The number that the *root* is calculated for is inside the radical symbol. The word radical is derived from the Latin word *rad,* which means root.

• The number inside the radical symbol is called the *radicand*. In the case of $\sqrt{x}$, x represents the radicand.

• Different *roots* of numbers exist. For example, for a number represented by x, possible roots are the square root, cubed root, fourth root, etc., represented by $\sqrt[2]{x}$, $\sqrt[3]{x}$, $\sqrt[4]{x}$, etc., respectively.

• An equivalent form of writing roots is $x^{1/2}$, $x^{1/3}$, $x^{1/4}$, where x represents any number. For example:

$$\sqrt[2]{x} = x^{1/2}$$

$$\sqrt[n]{x} = x^{1/n}$$

$$\sqrt[2]{45x^2} = (45x^2)^{1/2}$$

9.2 The Square Root

• The square root of a number, is equal to a number that, when squared, equals the original number.

• For example, the square root of 4 is equal to 2, such that 2^2 equals 4. Also, $(-2)^2$ is equal to 4, so the square root of 4 could also be –2.

• If a number has two equal factors, then it has a square root. For example, 4 has factors 2×2.

• A *perfect square* is a number that can be expressed as the product of two equal numbers or factors. The following are perfect squares:

1, 4, 9, 16, 25,

because they can be expressed as:

$1 \times 1, 2 \times 2, 3 \times 3, 4 \times 4$, and 5×5 respectively.

• The square root of 25 is ±5, because $5^2 = (5)(5)$ or $(-5)(-5) = 25$.

• The *radical symbol,* $\sqrt{\ }$, represents the positive square root.

• It is not possible to find a perfect square root of a negative number such as $\sqrt{-4}$ because when two identical numbers are multiplied together the result is always a positive number.

$2 \times 2 = 4$ and $-2 \times -2 = 4$

(Note: See Section 1.17, "Complex Numbers," for a discussion of $\sqrt{-1}$.)

• $\sqrt[2]{x}$ is commonly written $\sqrt{x}$, with the "2" not displayed.

• The following are examples of square roots:

$$\sqrt{25} = \sqrt{5 \times 5} = \sqrt{5^2} = \pm 5$$

$$\sqrt{36} = \sqrt{6 \times 6} = \sqrt{6^2} = \pm 6$$

$$\sqrt{9x^2} = \sqrt{(3x)(3x)} = \sqrt{(3x)^2} = \pm 3x$$

$$\sqrt{a^2} = \sqrt{(a)(a)} = \sqrt{(a)^2} = \pm a$$

$$\sqrt{0} = 0$$

$$\sqrt{1} = \pm 1$$

9.3 Estimating Square Roots

- To determine the square root of a number that is *not* a *perfect square*:

1. Determine what two perfect squares the number is between.
2. Estimate the square root of the number to one decimal point accuracy (0.1).
3. Divide your estimate into the number (and evaluate to the nearest thousandth).
4. Find the average of the estimate and the quotient.
5. Check how good the estimate is by squaring the result.

- A technical calculator can also be used to estimate square roots.

- **Example:** Estimate the square root of 12.

The two perfect squares this number is between are 9 and 16.

Therefore, the $\sqrt{12}$ is between $\sqrt{9}$ and $\sqrt{16}$.

Because $\sqrt{9} = 3$ and $\sqrt{16} = 4$, the $\sqrt{12}$ must be between 3 and 4.

Estimate the square root of the number to one decimal point accuracy (0.1)

Because 12 is 3 away from 9 and 4 away from 16, the estimate is 3.4.

Divide the estimate into 12 and evaluate it to the nearest thousandth.

Using a calculator $12 \div 3.4 = 3.529$.

Alternatively, using long division:

$$3.4\overline{)\,12}$$

Move the decimal point to the right in both numbers so there is no decimal in the divisor.

$$34.\overline{)\,120.00}$$

34 will divide into 120 three times. Multiply 3 by 34 and subtract the result from 120.

$$\begin{array}{r} 3.???\\ 34.\overline{)\,120.000}\\ \underline{102}\\ 18.0 \end{array}$$

Bring down the zero, divide 34 into 18.0, multiply the result by 34, and subtract that result from 18.0.

```
        3.5??
34.) 120.000
     102
       18.0
       17.0
         1.00
```

Bring down the zero, divide 34 into 1.00, multiply the result by 34, and subtract that result from 1.00.

```
        3.52?
34.) 120.000
     102
       18.0
       17.0
        1.00
         .68
         .32
```

Bring down the zero, and divide 34 into 0.320.

```
        3.529
34.) 120.000
     102
       18.0
       17.0
        1.00
         .68
         .320
         .306
         .014
```

Find the average of the estimate and the quotient.

$$\frac{3.4 + 3.529}{2} = \frac{6.929}{2} = 3.47$$

Check how good the result is by squaring it:

$$(3.47)^2 = 12.04.$$

This process can be repeated to better the approximation by dividing 3.47 into 12 and proceeding as above.

9.4 The Cubed Root

- The cubed root of a positive number is the number that, when multiplied by itself three times, equals that number.

- **Example:** The cubed root of 8 written, $\sqrt[3]{8}$ is equal to 2.

This is true because, $2^3 = (2)(2)(2) = 8$

Also:

$$\sqrt[3]{8} = \sqrt[3]{(2)(2)(2)} = \sqrt[3]{2^3} = 2$$

If *a* represents a number then:

$$\sqrt[3]{(a)(a)(a)} = \sqrt[3]{a^3} = a$$

9.5 The Fourth and Fifth Roots

• The fourth root of a positive number equals a number that when multiplied by itself four times equals the number.

• **Example:**

$$\sqrt[4]{16} = \sqrt[4]{(2)(2)(2)(2)} = \sqrt[4]{2^4} = 2$$

$$\sqrt[4]{a^4} = \sqrt[4]{(a)(a)(a)(a)} = a$$

Where *a* represents a number.

• The fifth root of a positive number equals a number that when multiplied by itself five times equals the number.

• **Example:**

$$\sqrt[5]{32} = \sqrt[5]{(2)(2)(2)(2)(2)} = \sqrt[5]{2^5} = 2$$

9.6 Simplifying Radicals by Factoring

• To simplify radicals by factoring, factor the radicand, look for perfect squares for square roots, or cubes for cubed roots, etc., and if multiple factors exist, bring each root out of the radical.

• **Example:** Simplify the following by factoring and reducing the radicals.

$$\sqrt{36} = \sqrt{6 \times 6} = 6$$

$$\sqrt{32} = \sqrt{8 \times 4} = \sqrt{2 \times 2 \times 2 \times 2 \times 2} = 2 \times 2\sqrt{2} = 4\sqrt{2}$$

$$\sqrt{42} = \sqrt{6 \times 7} = \sqrt{2 \times 3 \times 7}$$

There are no squared factors, so this radical cannot be reduced further.

$$\sqrt{a^2b^3} = \sqrt{aabbb} = ab\sqrt{b}$$

$$\sqrt{45x^5} = \sqrt{(9)(5)(x^2)(x^2)(x)}$$

$$= \sqrt{(3^2)(x^2)(x^2)(5)(x)} = 3(x)(x)\sqrt{(5)(x)} = 3x^2\sqrt{5x}$$

$$\sqrt[3]{8} = \sqrt[3]{(2)(2)(2)} = \sqrt[3]{2^3} = 2$$

$$\sqrt[4]{16} = \sqrt[4]{(2)(2)(2)(2)} = \sqrt[4]{2^4} = 2$$

9.7 Multiplying Radicals

- To multiply radicals, combine the numbers under one radical symbol, multiply, and simplify.

- **Example:** Multiply the following radicals:

$$\sqrt{3}\sqrt{6} = \sqrt{3 \times 6} = \sqrt{3 \times 3 \times 2} = 3\sqrt{2}$$

$$\sqrt{a}\sqrt{b} = \sqrt{ab}$$

9.8 Dividing Radicals

- To divide radicals, combine the numbers under one radical symbol, divide, and simplify.

- **Example:** Divide the following radicals:

$$\sqrt{3} \div \sqrt{6} = \frac{\sqrt{3}}{\sqrt{6}} = \sqrt{\frac{3}{6}} = \sqrt{\frac{3}{3 \times 2}} = \sqrt{\frac{1}{2}} = \frac{1}{\sqrt{2}}$$

(Note that the 3 in the numerator canceled with the 3 in the denominator.)

$$\sqrt{8} \div \sqrt{6} = \frac{\sqrt{8}}{\sqrt{6}} = \sqrt{\frac{8}{6}} = \sqrt{\frac{(2)(2)(2)}{(3)(2)}} = (2)\sqrt{\frac{1}{3}} = \frac{2}{\sqrt{3}}$$

$$\sqrt{a} \div \sqrt{b} = \frac{\sqrt{a}}{\sqrt{b}} = \sqrt{\frac{a}{b}}$$

9.9 Radicals Involving Fractions

• To simplify a radical containing fractions or a fraction containing radicals, combine the radicands in each radical under one radical symbol, divide, and simplify.

• **Example:** Simplify the following radicals:

$$\frac{\sqrt{a}}{\sqrt{b}} = \sqrt{\frac{a}{b}}$$

$$\sqrt{\frac{9}{36}} = \sqrt{\frac{3 \times 3}{6 \times 6}} = \frac{3}{6} = \frac{1}{2}$$

9.10 Rationalizing the Denominator

• Rationalizing the denominator of a fraction containing radicals creates a fraction without any radicals in the denominator.

• To rationalize the denominator, simplify the radical in the denominator, multiply both the numerator and the denominator by the radical in the denominator (that will make the denominator's radicand a perfect square), then simplify the fraction.

• **Example:** Rationalize the denominator for $1/\sqrt{8}$.

First simplify the denominator.

$$\frac{1}{\sqrt{8}} = \frac{1}{\sqrt{2\times2\times2}} = \frac{1}{2\sqrt{2}}$$

Multiply the numerator and denominator by $\sqrt{2}$.

$$\frac{(1)\sqrt{2}}{2\sqrt{2}\sqrt{2}} = \frac{\sqrt{2}}{2\sqrt{2\times2}} = \frac{\sqrt{2}}{2\times2} = \frac{\sqrt{2}}{4}$$

9.11 Addition of Radicals

• To add radicals, express each radical individually, then add. The radicands cannot be combined under one radical symbol.

• **Example:** Add the following radicals:

$$\sqrt{8}+\sqrt{12} = 2\sqrt{2}+2\sqrt{3}$$

$$\sqrt{4}+\sqrt{6} = 2+\sqrt{6}$$

$$\sqrt{36}+\sqrt{25} = 6+5 = 11$$

$$3\sqrt{6}+4\sqrt{6}+5\sqrt{6} = (3+4+5)\sqrt{6} = 12\sqrt{6}$$

$$3\sqrt{5}+4\sqrt{6}+5\sqrt{5}+7\sqrt{6}$$

$$= (3+5)\sqrt{5}+(4+7)\sqrt{6} = 8\sqrt{5}+11\sqrt{6}$$

(Note that $\sqrt{6}$ and $\sqrt{5}$ were factored out of each term in the above two examples.)

- Adding and subtracting radicals is similar to combining like terms in algebraic equations.

9.12 Subtraction of Radicals

- To subtract radicals, express each radical individually, then subtract. The radicands cannot be combined under one radical symbol.

- **Example:** Subtract the following radicals:

$$\sqrt{36} - \sqrt{25} = 6 - 5 = 1$$

$$\sqrt{4} - \sqrt{6} = 2 - \sqrt{6}$$

$$5\sqrt{6} - 4\sqrt{6} = (5 - 4)\sqrt{6} = 1\sqrt{6} = \sqrt{6}$$

$$7\sqrt{6} - 4\sqrt{6} - 5\sqrt{5} = (7 - 4)\sqrt{6} - 5\sqrt{5} = 3\sqrt{6} - 5\sqrt{5}$$

Notes

Chapter 10

Important Statistical Quantities

10.1 Average (Arithmetic Mean)

In this section the average or arithmetic mean and the weighted average are defined.

- The *average* and the *arithmetic mean* are equivalent.
- To find the average of a group of numbers, add all of the numbers together and divide the sum by how many numbers there are.

$$\text{Average} = \frac{\text{total sum of the numbers}}{\text{how many numbers there are}}$$

- **Example:** What is the average of $\{5, 6, 9, 25\}$?

The total sum is:

$$5 + 6 + 9 + 25 = 45$$

Divide the sum by how many numbers there are, or 4.

$$45 \div 4 = 11.25$$

- **Example:** What is the average of $\{9, 6, 6, 9, 25, 25\}$?

The total sum is:

$$9 + 6 + 6 + 9 + 25 + 25 = 80$$

Divide the sum by how many numbers there are, or 6.

$$80 \div 6 = 13.333333....$$

Which rounds to 13.33.

• To solve a problem using the *weighted average,* each quantity is weighted (multiplied) by the number of items having that quantity. After multiplying each quantity by the number of items having that quantity, add the products and divide the sum by the total number of items.

• For example, what is the weighted average of the following: If you purchase 5 shrubs and two cost \$2 and three cost \$7, what is the average cost per shrub?

To solve this problem as a weighted average, the two costs are weighted (multiplied) by the number of items and the products are added.

$2(\$2) + 3(\$7) = \$25$

$\$4 + \$21 = \$25$

Then, divide the sum by the total number of shrubs.

$\$25 \div 5 = \$5$

Therefore, the average cost per shrub is \$5.

This problem can also be solved as a standard average.

The total cost is:

$\$2 + \$2 + \$7 + \$7 + \$7 = \$25$

Divide the sum by the total number of shrubs, or 5.

$\$25 \div 5 + \$5$

Therefore, the average cost per shrub is \$5.

10.2 Median

- The median is the value of the number in the middle of a sequential list of numbers. If there are two numbers in the middle, the median is the value of their average. There must be an equal number of numbers above and below the median.

- To find the median of a group of numbers, list the numbers sequentially; the median is either the number in the middle, or the average of the two numbers in the middle.

- To find the median value of an odd number of numbers, list the numbers in an ordered sequence from least in value to greatest; the median is the number in the middle.

- **Example:**

  Find median of $\{5, 7, 3, 10, 1\}$.

  List in order. $1, 3, 5, 7, 10$

  The middle number, or median, is 5.

- To find the median value of an even number of numbers, list the numbers in an ordered sequence from least in value to greatest; the median is the average of the two middle numbers.

- **Example:**

  Find median of $\{5, 7, 3, 12, 10, 1\}$

  List in order. $1, 3, 5, 7, 10, 12$

  The middle two numbers are 5 and 7.

The average of the middle two numbers is:

$$\frac{5+7}{2} = \frac{12}{2} = 6$$

Therefore, the median is 6.

10.3 Mode

- The mode of a list of numbers is the number that occurs most frequently.

- **Example:** Find the mode of:

$$\{3, 6, 29, 8, 6, 4, 3, 85, 2, 8, 35, 3, 8, 9\}.$$

It is easier to list the numbers sequentially.

$$2, 3, 3, 3, 4, 6, 6, 8, 8, 8, 9, 29, 35, 85$$

The numbers 3 and 8 both occur three times, which is the greatest number of times.

Therefore, 3 and 8 are both modes.

- As shown in this example, there can be more than one mode.

- If no number occurs more than once, then the mode does not exist for that list.

10.4 Probability

- Probability defines the chances of some certain event occurring.

- The probability, p, is generally expressed as a fraction.

$$\text{Probability} = \frac{\text{number of possible ways}}{\text{total number of ways}}$$

- For example, if a cube has one yellow face and five blue faces, what is the probability that the yellow face will land on top if the cube is tossed in the air?

The answer is: $p = 1/6$

10.5 Standard Deviation, Variance, Histograms, and Distributions

In this section, standard deviation, variance, histograms, and distributions are defined and explained.

Standard Deviation and Variance

- Consider experimental data that has been collected where a measurement, x_1, is made of a quantity x. If other measurements are made, we expect random error within our measurement that will be distributed around the correct value. The standard deviation and the variance represent the uncertainties associated with experimental attempts to determine the true value of some quantity.

- The symbol for standard deviation is sigma σ and the symbol for variance is σ^2.

- The *standard deviation* is a measure of the dispersion in a frequency distribution. It is equal to the square root of the mean of the squares of the deviations from the arithmetic mean of the distribution. The *variance* is the square of the standard deviation. The standard deviation is the square root of the variance.

The standard deviation: $\sigma = \pm\sqrt{\sigma^2}$

Where σ^2 is the variance.

$$\sigma^2 = \text{Lim}_{N->\infty}\left[\frac{1}{N}\sum(x_i - \mu)^2\right]$$

$$= \text{Lim}_{N->\infty}\left(\frac{1}{N}\sum x_i^2\right) - \mu^2 = \langle x^2 \rangle - \mu^2$$

Where:

$<x>$ is a weighted average over x of all possible values of x.

x is the quantity that is being measured.

x_i represents the measurements made.

N is the number of measurements made, and i=1 is the first measurement.

The sum (Σ) of all measurements from i=1 to i=N is given by:

$$\sum_{i=1}^{N}(x_i) = x_1 + x_2 + x_3 + x_4 + \ldots + x_N = \sum(x_i)$$

The $\mathrm{Lim}_{N->\infty}$ represents the Limit as the number of measurements get larger and larger, (see Chapter 5, "Limits," in *Master Math Pre-Calculus*). μ represents the average of all possible measurements.

$$\mu = \mathrm{Lim}_{N->\infty}\left(\frac{1}{N}\sum x_i\right)$$

The average is the sum of the values of the measurements divided by the total number of measurements, N.

- The variance (and from that the standard deviation) can be calculated using the equation:

$$\sigma^2 \sim \frac{1}{N-1}\left[\sum x_i^2 - (x_{ave})\sum x_i\right]$$

$$= \frac{1}{N-1}\left[\sum x_i^2 - N(x_{ave})^2\right]$$

x_{ave} represents the average of the measurements (the experimental mean).

The Histogram, Binomial Distribution, Gaussian Distribution, and Poisson Distribution

• If 100 measurements of the length of a football field are made, the measurements can be graphed in terms of a frequency distribution indicating the number of times each length was measured. A mean of the data points can be calculated and a standard deviation estimated. A *histogram graph* is used to depict the number of measurements recorded for each length.

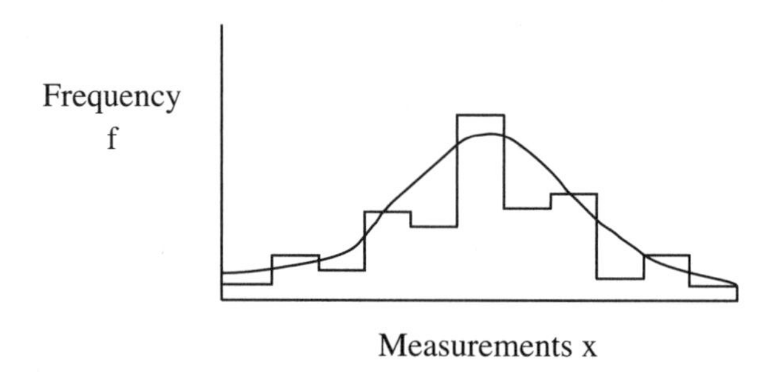

The curve represents an estimate distribution assuming infinite measurements could be made.

• The *binomial distribution* describes the probability of observing x for n attempts where only two mutually exclusive outcomes are possible. The probability for success in each attempt is p. A *binomial experiment* is an experiment with n repeated attempts where: a) the attempts are independent; b) each attempt results in only two possible outcomes, "success" and "failure"; and c) the probability, p, of a success on each attempt, remains constant. x is the number of attempts that result in success and has a binomial distribution with parameters p and n = 1, 2, 3, ... The probability function is:

$$P_B = \frac{n!}{x!\,(n-x)!} p^x (1-p)^{n-x}$$

Where $\mu = np$ and $\sigma^2 = np(1-p)$.

(μ represents the mean and σ represents the standard deviation for the distribution of the hypothetical infinite population of all possible observations.)

The total number of different sequences of attempts that contain x successes and n – x failures is:

$$\frac{n!}{x!\,(n-x)!}$$

• The *Gaussian distribution* or normal error distribution, is a bell-shaped curve and describes an expected distribution of random observations for a given experiment. In fact, it seems to describe the distribution of estimations of the parameters of most probability distributions. The Gaussian distribution is an approximation of the binomial distribution for the case where the number of possible different observations is large and the probability of success for each measurement is large. The Gaussian distribution is useful for smooth symmetrical distributions with large n and infinite p.

$$P_G = \frac{1}{\sigma\sqrt{2\pi}}\exp\left[-\frac{1}{2}\left(\frac{x-\mu}{\sigma}\right)^2\right]$$

(Note that exp means raise to the exponent e.)

• The *Poisson distribution* represents an approximation to the binomial distribution for the case when the average number of successes is smaller than the possible number. The Poisson distribution is useful for describing small samples with large numbers of events that are difficult to observe.

$$P_P = \frac{\mu^x}{x!}\exp[-\mu]$$

For $\sigma^2 = \mu$

Notes

Index

D

Q

R

W

Z